BIOELECTRONIC VISION
Retina Models, Evaluation Metrics, and System Design

SERIES ON BIOENGINEERING AND BIOMEDICAL ENGINEERING

Series Editor: John K-J Li *(Department of Biomedical Engineering, Rutgers University, USA)*

Vol. 1:   Dynamics of the Vascular System
        by John K-J Li *(Department of Biomedical Engineering, Rutgers University, USA)*

Vol. 2:   Neuroprosthetics
        Theory and Practice
        eds. Kenneth W Horch and Gurpreet S Dhillon *(Department of Bioengineering, University of Utah, USA)*

Vol. 3:   Bioelectronic Vision
        Retina Models, Evaluation Metrics, and System Design
        by João Carlos Martins *(ESTIG-IPBeja/INESC-ID, Portugal)* and Leonel Augusto Sousa *(IST-TULisbon/INESC-ID, Portugal)*

Vol. 4:   Understanding the Human Machine
        A Primer for Bioengineering
        by Max E Valentinuzzi *(Universidad Nacional de Tucumán, Argentina)*

Series on Bioengineering & Biomedical Engineering – Vol. 3

BIOELECTRONIC VISION
Retina Models, Evaluation Metrics, and System Design

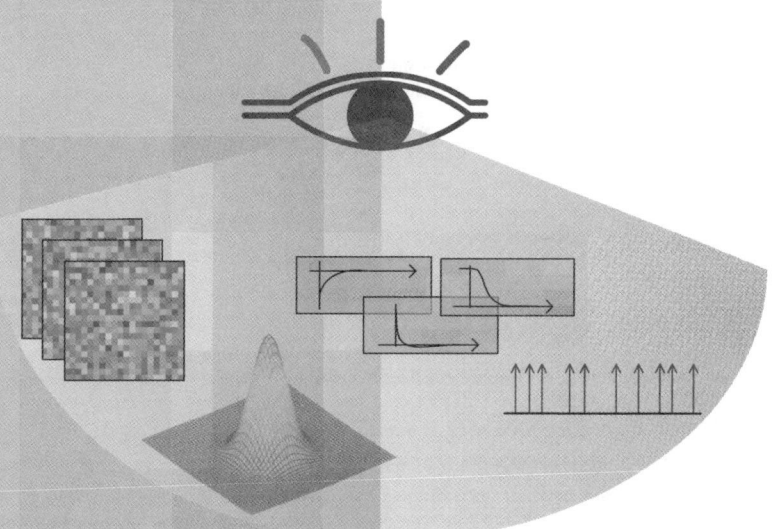

João C. Martins
ESTIG-IPBeja/INESC-ID, Portugal

Leonel A. Sousa
IST-TULisbon/INESC-ID, Portugal

World Scientific

NEW JERSEY · LONDON · SINGAPORE · BEIJING · SHANGHAI · HONG KONG · TAIPEI · CHENNAI

Published by

World Scientific Publishing Co. Pte. Ltd.
5 Toh Tuck Link, Singapore 596224
USA office: 27 Warren Street, Suite 401-402, Hackensack, NJ 07601
UK office: 57 Shelton Street, Covent Garden, London WC2H 9HE

British Library Cataloguing-in-Publication Data
A catalogue record for this book is available from the British Library.

Series on Bioengineering & Biomedical Engineering — Vol. 3
BIOELECTRONIC VISION
Retina Models, Evaluation Metrics, and System Design

Copyright © 2009 by World Scientific Publishing Co. Pte. Ltd.

All rights reserved. This book, or parts thereof, may not be reproduced in any form or by any means, electronic or mechanical, including photocopying, recording or any information storage and retrieval system now known or to be invented, without written permission from the Publisher.

For photocopying of material in this volume, please pay a copying fee through the Copyright Clearance Center, Inc., 222 Rosewood Drive, Danvers, MA 01923, USA. In this case permission to photocopy is not required from the publisher.

ISBN-13 978-981-279-430-7
ISBN-10 981-279-430-1

Printed in Singapore by Mainland Press Pte Ltd

To our families

Preface

Vision is one of the most important senses for the human being. Blindness can be a serious impairment, preventing a person from having an ordinary life. It restricts one of the most important streams of information from the world, thus limiting any communication made through visual expression and visual arts. What is commonly understood as vision relies on a very complex system integrating the eyes, the optical nerve and the brain. Research continues in order to gain a better understanding of the structure and operation of the visual system components: the retina, the lateral geniculate nucleus, and the visual cortex. The retina is a crucial layer in the vision process, while the visual cortex is the main brain visual processing area. One of the final goals, and the most expected result of the current research effort, is the development of artificial vision systems that can restore vision to blind people. The latest advances in electronics and biology have opened new possibilities that will finally allow bio-inspired electronic devices and systems to be designed that partially replace damaged parts of the visual system. This challenging domain, which gathers knowledge from distinct disciplines, including biology, neurology, and engineering, is called *BioElectronic Vision*.

A fully functional bioelectronic vision system that provides vision restoration to completely blind people is not yet available. Since the road ahead is mainly unknown, we have to increase current knowledge and improve practices in small steps. One of the required steps is the writing of books and monographs that present, synthesize and analyze current knowledge in all the distinct domains of bioelectronic vision. This work is a contribution towards this goal. The central goals of this book are: *i)* to establish a background for an engineering audience on the physiological constitution of the human visual system; *ii)* to present the different classes of computational retina models, *iii*) to analyze the neural activity metrics; and finally, *iv*) to discuss the electronic devices and circuits required to implement a bioelectronic vision system and metrics to determine how well it accomplishes its goal.

This book is intended to be used as a reference for those involved in the area of bioelectronic vision, in particular in modeling the retina behavior and in the design of embedded systems that underlie visual neuroprostheses. It can also be used as a textbook for first-year graduate or senior courses on advanced signal processing for artificial vision. Codification of the information in visual systems and visual prostheses are frequently topics found in bioengineering and electrical engineering curricula. We assume the reader has a basic knowledge of signal processing and programming, as might be learned with some job experience or as part of a first undergraduate course in these topics.

While keeping the goal in mind - the implementation of an electronic visual prosthesis for bioelectronic vision - the anatomy and operation of the retina as well as the main features of the neuronal responses are presented. Retina models, representative of different model classes, are presented and a framework for training, evaluating and comparing them is established. This framework includes a set of different metrics that allows the testing and evaluation of not only the actual models, but also new retina models that will almost certainly be proposed in the future. The book follows a digital signal processing perspective, where all models and metrics are discretized, implemented, and tested in a computer environment. These computational models and programs can be implemented in an embedded system specifically designed to be an electronic visual prosthesis. The final chapter of this book is devoted to the technological challenges needed to adapt these embedded systems into artificial visual systems that can thus be used to restore visual sense to blind people.

This book was prepared to be read from cover to cover. However, while reading each chapter in order, one can skip a chapter depending on the reader's background and the chapter's purpose. Chapter 1 provides an overview of the topics related to bioelectronic vision, namely the building blocks of the visual system, the characteristics and causes of blindness, and the main components of a neuroprosthesis. Chapter 2 describes the human visual system and identifies the main classes of retina models. (This chapter can be skipped if the reader has a good background in the anatomy and operation of the human visual system.) Chapter 3 is devoted to the establishment of the required background in order to analyze neural systems in general, and the retinal neural code in particular. Chapter 4 is dedicated to the description and analysis of the considered functional and structural classes of retina models. Chapter 5 provides a set of collected neural activity metrics, which are used throughout the book to train, test and evaluate the retina models. This chapter also presents experimental data and gives some insights into the metric's performance and its drawbacks. Chapter 6 concludes the book by presenting the main electronic components required to prototype a bioelectronic system, while discussing the principal technological features of its implementation. Chapter 5 and Chap. 6 are chiefly important for people interested in the implementation of artificial retinas and visual neuroprostheses. Each book chapter is accompanied by a group of exercises that allows the reader to evaluate his/her knowledge while guiding the reader through the implementation of the methods, algorithms and techniques presented in the book. These exercises can be a valuable instrument for instructors using the book for teaching.

The participation of the authors during the last years in research projects to develop Bioelectronic Vision systems has given them the opportunity to have a central perspective on the research in this area – namely the project "Cortical Visual Prosthesis for the Blind" (CORTIVIS), supported by the Commission of the European Communities, and the "Retinal Neural Code: Accurate Modeling Toward an Artificial Visual System" (RNC) project, supported by the Portuguese Foundation for Science and Technology (FCT). We did our best to provide both solid theoretical support and all the information needed to apply this knowledge to the design of bioelectronic vision systems. Our main goal is to contribute to the education of the next generation of researchers in this new multidisciplinary area of bioelectronic vision. We really hope you find the book useful!

João C. Martins
Leonel A. Sousa

Acknowledgments

This book is not by any means an individual work, but instead it is a reflection of different contributions from many people.

We want to express our gratitude to the members of the Signal Processing Systems (SIPS) research group of the Instituto de Engenharia de Sistemas e Computadores - Investigação e Desenvolvimento em Lisboa INESC-ID for the friendly and challenging work environment. In particular, we want to thank Pedro Tomás for the fruitful discussions and suggestions, José Germano for implementing the retina models in the FPGA, and Moisés Piedade for the electronics necessary to implement the wireless link and the stimulator in the presented prototype.

We would like also to express our appreciation to Eduardo Fernández and Markus Bongard from the Instituto de Bioingenieria/Universidad Miguel Hernández, Elche, Spain, who guided us through the complexity and beauty of the human visual system. The lab facilities at their institute have been fundamental in gathering retina data, like that presented in this book. Our thanks, too, to Markus Meister for providing the Gaussian flicker data used in this book.

We also acknowledge the institutions that supported this work. First, we would like to acknowledge INESC-ID for supporting our research continuously throughout the last years. We would like also to thank those who financially supported the work behind this book, the Commission of the European Communities under the CORTIVIS contract (QLK6-CT-2001-00279), and the Portuguese Fundação para a Ciência e Tecnologia (FCT), namely under the project RNC (POSI/EEA-CPS/61779/2004).

Last, but not least, we would like to express our gratitude to our families for their constant and invaluable support.

Contents

Preface   vii

Acknowledgments   xi

List of Figures   xvii

List of Tables   xxi

List of Acronyms   xxiii

1. Introduction to Bioelectronic Vision   1
    1.1 Main Causes of Blindness   2
    1.2 Main Components of a Bioelectronic Vision System   5
    1.3 Classification of Visual Prostheses   10
        1.3.1 Retinal Neuroprosthesis   12
        1.3.2 Cortical Visual Neuroprosthesis   14
    1.4 Conclusions and Further Reading   15

2. The Human Visual System   21
    2.1 Introduction   21
    2.2 The Neuron   22
        2.2.1 Neuron Anatomy   22
        2.2.2 Neuron Dynamics   23
    2.3 The Human Visual System   28
        2.3.1 The Eye   29
        2.3.2 The Retina   31
        2.3.3 How the Retina Operates   37
        2.3.4 The Visual Pathway   44

	2.4	Modeling the Retina	49
		2.4.1 The Retina Neural Code	49
		2.4.2 Classification of Retina Models	52
	2.5	Conclusions and Further Reading	54
3.	Characterization of the Neural Response		57
	3.1	Introduction	57
	3.2	Spikes: The Essence of the Neural Code	58
		3.2.1 Retina Stimulation and Responses Recording	60
		3.2.2 Spike Trains and Firing Rates	71
		3.2.3 Spike Triggered Average	83
		3.2.4 Spike Train Autocorrelation Function	87
		3.2.5 The Spike Triggered Covariance	89
	3.3	Stimulus and Response Statistics, and Firing Probabilities	90
		3.3.1 Spike Train Statistics	92
		3.3.2 Homogeneous Poisson Model of Spike Trains	93
		3.3.3 Inhomogeneous Poisson Model of Spike Trains	100
		3.3.4 Spike-Count Statistics	102
	3.4	Spiking Mechanisms	103
		3.4.1 Generation of Poisson Spike Trains	104
		3.4.2 Integrate-and-Fire Spike Generation	105
	3.5	Conclusions and Further Reading	107
4.	Retina Models		113
	4.1	Introduction	113
	4.2	Classification of Retina Models	113
	4.3	Structural Models	115
		4.3.1 The Integrate and Fire Model	115
		4.3.2 The Leaky Integrate-and-Fire Model	118
	4.4	Functional Models	123
		4.4.1 Deterministic Models	124
		4.4.2 Stochastic Models	129
		4.4.3 White Noise based Models	138
	4.5	Conclusions and Further Reading	145
5.	Neural Activity Metrics and Models Assessment		155
	5.1	Introduction	155
	5.2	The Metric Definition	155

5.3	Firing Rate Metrics		156
	5.3.1	Mean Squared Error	156
	5.3.2	Normalized Mean Squared Error	157
	5.3.3	Percent Variance Accounted For	158
	5.3.4	Analysis of the Firing Rate Metrics	159
5.4	Spike Train Metrics		160
	5.4.1	Spike Time Metric	161
	5.4.2	Interspike Interval Metric	166
	5.4.3	Spike Train Distance Metric	170
	5.4.4	Spike Train Metrics Analysis	174
5.5	Spike Events Metrics		177
	5.5.1	Spike Events Metric Analysis	187
5.6	Tuning and Assessment of Retina Models		188
	5.6.1	Deterministic Model	189
	5.6.2	Stochastic Model	191
	5.6.3	White Noise Model	191
5.7	Conclusions and Further Reading		193

6. **Design and Implementation of Bioelectronic Vision Systems**   199

6.1	Retinal Prostheses		199
	6.1.1	Epiretinal Implants	200
	6.1.2	Subretinal Implants	202
6.2	Retinal Bioelectronic Vision System Design		204
6.3	Cortical Visual Prostheses		208
6.4	Cortical Bioelectronic Vision System Design		212
	6.4.1	Early Layers	213
	6.4.2	Neuromorphic Pulse Coding	217
	6.4.3	Spike Multiplexing	218
	6.4.4	Serial Communication Link	221
6.5	Vision Prosthesis Prototype		223
6.6	Conclusions and Further Reading		226

Bibliography   233

Index   241

List of Figures

1.1	Causes of blindness in the world.	3
1.2	World map of vision neuroprosthesis groups.	6
1.3	Diagram of the human visual system.	7
1.4	Main components of a visual neuroprosthesis.	8
1.5	Main components of a retinal neuroprosthesis	13
2.1	Neuron anatomy.	24
2.2	Propagation of an action potential along a neuron's axon.	25
2.3	Neuron gap junction.	28
2.4	Diagram section of the human eye.	29
2.5	Light micrograph of a vertical section through the retina.	32
2.6	Simplified schematic organization of the retina.	33
2.7	Photoreceptors' spectral sensitivity.	34
2.8	Spatial distribution of photoreceptors.	34
2.9	Photographs of the cross section of the fovea and of the foveal periphery.	35
2.10	Human retina photograph.	36
2.11	The receptive field of a cone.	38
2.12	Photoreceptor to bipolar cell connections.	39
2.13	Connections of a horizontal cell.	41
2.14	Connections between bipolar cells and ganglion cells.	42
2.15	AII amacrine cell connections.	44
2.16	Visual pathways	46
2.17	Lateral geniculate nucleus cell layers	47
3.1	Representation of a spike train.	58
3.2	Spike waveform of a ganglion cell from a rabbit's retina.	59
3.3	Neuronal response function of a retinal ganglion cell.	59

3.4	Cellular recording of neuronal signals.	61
3.5	Rabbit type-ON RGC responses for a ON-OFF full-field stimulus.	63
3.6	Salamander type-ON RGC responses for sampled white-noise full-field stimulus.	64
3.7	Spatially uniform visual stimuli.	65
3.8	Spatially non-uniform visual stimuli.	65
3.9	Spatially nonuniform Gabor functions.	66
3.10	Stimuli with spatial and temporal modulation.	67
3.11	Gaussian white noise stimulus sequence with spatial, temporal and chromatic variation.	68
3.12	Spatially non-uniform visual stimuli	69
3.13	Microelectrode array.	69
3.14	Experimental apparatus for retina data acquisition and analysis.	70
3.15	Retina preparation for data acquisition.	71
3.16	The δ_Δ function.	73
3.17	Neural spike trains from a Salamander ON-type retinal ganglion cell and the stimulus	74
3.18	Firing rate ON-type RGC.	75
3.19	The rectangular (boxcar) filter window.	79
3.20	The Gaussian filter window.	80
3.21	The α function filter.	81
3.22	Firing rate obtained by filtering the neural response with different types of filter windows.	82
3.23	Procedure for the STA computation.	84
3.24	Spike triggered average time reversed	85
3.25	Spike number probability density for a train described by a homogeneous Poisson process.	95
3.26	Interspike time interval exponential probability density for a spike train described by a homogeneous Poisson process.	95
3.27	Integrate-and-fire spike generation from firing rate.	106
4.1	The leaky integrate-and-fire (LI&F) model.	116
4.2	Integrate and fire model's responses for constant input stimulus current	117
4.3	Leaky integrate-and-fire model's response for constant input stimulus current	120
4.4	Firing rate versus the stimulus current for integrate-and-fire models.	122
4.5	Integrate-and-fire model of the retina.	123

4.6	Block diagram of the deterministic model.	124
4.7	Discrete spatial DoG.	128
4.8	Block diagram of the pseudo-stochastic model.	130
4.9	Distorted sinus basis functions.	133
4.10	Spike triggered average represented by bases functions.	134
4.11	Generation of a noise sequence with a specific autocorrelation.	137
4.12	The white noise model structure	138
4.13	White noise model characterization of a salamander and of a rabbit RGC.	144
4.14	Function integral between nT_s and $(n+1)T_s$.	149
4.15	Forward rectangular approximation for the integral.	149
4.16	Trapezoidal approximation for the integral.	149
4.17	Backward approximation for the integral.	149
5.1	Comparison of two firing rates.	157
5.2	Comparing smoothed and not smoothed PSTHs.	161
5.3	Comparison of two spike trains.	161
5.4	Path to transform a spike train into the other in the spike time metric.	163
5.5	Path to transform a spike train into another in the interspike interval metric.	167
5.6	Changing of the spike shape for comparison with the spike distance metric.	172
5.7	Evolution of the spike time and interspike metrics for a set of spike trains for a salamander ON-type RGC.	175
5.8	Distance metric between a salamander ON-type RGC responses.	177
5.9	Parsing a set of spike trains into firing events.	178
5.10	Characterization of spike trains into firing events.	181
5.11	Deterministic model responses, when tuned with the NMSE and with the spike events metrics, to the flash stimulus.	190
5.12	Deterministic model responses, when tuned with the NMSE and with the spike events metrics, to the white noise stimulus.	190
5.13	Stochastic model responses when tuned with the NMSE and with the spike events metrics to the flash stimulus.	191
5.14	Responses of the stochastic model to the white noise stimulus, when tuned with the NMSE and spike event metrics.	192
5.15	White noise model responses to the flash and to the white noise stimuli.	192
5.16	Models errors for the ON-OFF stimulus.	193

5.17 Models errors for the white noise stimulus. 194

6.1 IRP test device. 205
6.2 Layout of a retinal implant. 206
6.3 Electrode layout. 209
6.4 A diagram of the Dobelle apparatus. 209
6.5 Microelectrode array. 211
6.6 Bioelectronic vision system based on the CORTIVIS neuroprosthesis. 212
6.7 Modules of the bio-inspired processing module of the artificial retina. 212
6.8 Extended RGB model for replacing natural human retina processing. 213
6.9 Architecture for computing the spatial filter. 214
6.10 Signal flow graphs for the temporal (high pass filter and contrast gain control) and rectifier components for the *Early Layers* model. 215
6.11 Architecture of the temporal part of the *Early Layers* model. . 216
6.12 Simplified architecture of the *Neuromorphic Pulse Coding* module; one register per microelectrode. 217
6.13 Structure of an Address Event Representation (AER) tree and possible implementations for the arbiters. 219
6.14 Architecture of the FIFO based AER module. 220
6.15 Block diagram of the RF link and the *Microelectrode Stimulator* module. 221
6.16 The RF link. 223
6.17 Elonica prototype vision system. 224
6.18 Matlab code to read and write an AVI file. 230

List of Tables

1.1 Main pros and cons of visual prostheses approaches. 19
5.1 Values of the firing rate metrics applied to salamander ON-type RGC responses. 160
5.2 Limit values for the spike train metrics using the neuronal responses of a ON-type salamander RGC. 175
6.1 FPGA circuit area (from [Piedade et al. (2005)]). 225
6.2 Main features of implants for visual neuroprostheses. 226

List of Acronyms

AER Address Event Representation
AGC automatic gain control
AMD age-related macular degeneration
ASK Amplitude Shift Keying
BPSK Binary Phase Shift Keying
CCD charged couple device
CDF cumulative density function
CGC contrast gain control
CRT cathode ray tube
DAC digital-to-analog converter
DoG difference of Gaussians
EM Expectation-Maximization
FIFO First In First Out
FPGA Field Programable Gate Array
fps frames-per-second
FSK Frequency Shift Keying
GCL ganglion cell layer
I&F integrate-and-fire
INL inner nuclear layer
IPL inner plexiform layer
IRP Intraocular Retinal Prosthesis
ISI interspike time interval
LCD liquid crystal display
LI&F leaky integrate-and-fire
LGN lateral geniculate nucleus
LoG Laplacian of Gaussian
MEA microelectrode array

MEMS microelectromechanical systems
modDoG modified-difference of Gaussians
MSE mean squared error
NIR near-infrared
NMSE normalized mean squared error
ONL outer nuclear layer
OPL outer plexiform layer
pdf probability density function
pixel picture element
pmf probability mass function
PCA principal component analysis
PSG Poisson spike generator
PSTH peri-stimulus time histogram
RAM random access memory
RF receptive field
RGB red-green-blue
RGC retinal ganglion cell
ROC region of convergence
ROM read only memory
RP retinitis pigmentosa
SFG signal flow graph
STA spike triggered average
STC spike triggered covariance
TFT thin-film transistor
%VAF percent-Variance-Accounted-For
V1 cortex visual area 1
VLSI very large scale integration

Chapter 1

Introduction to Bioelectronic Vision

As everyone would certainly agree, vision is the most crucial sense; it provides the primary means of gathering information from the surrounding environment starting from birth. During our lifetimes, vision is the sense through which most of the information is perceived and, in many ways, is responsible for and affects who we are. For example, it is through vision that human beings perceive art. In everyday life, vision is an indispensable resource used in performing the most simple tasks. People who unfortunately lost their vision can become very dependent upon others, which in most cases represents a social problem.

Blindness is a severe impairment, since for the blind, it is very difficult to recognize other people, landscapes, and objects of daily life. In addition, they usually have severe motion restrictions, since they are depending on others to move safely. People who have been able to see for years lose an essential contributor to their quality of life due to blindness. Loss of vision is not only an enormous psychological burden, but it can also cause severe handicaps and cause tremendous difficulties in moving in strange, and even in formerly familiar, environments.

In 2002, it was estimated that more than 161 million people were visually impaired, of whom, 124 million people had low vision and 37 million were blind. However, refractive error as a cause of vision impairment was not included in these figures, which implies that the actual global magnitude of vision impairment is greater [World Health Organization (2004)].

Human vision disabilities are of different types and have different geneses. Thus, depending on the nature of the disability, different approaches have to be used to circumvent the dysfunction. Impairments in the eye's optical system, which is responsible for transmitting and focusing light as a sharp image on the retina, usually are easily overcome with the use of an external corrective optical system. Optical lenses (glasses) are a typical optical choice to remedy this problem; another option is surgical intervention, like in the case of cataracts, where an eye lens transplant can be performed.

The current challenge is to circumvent retinal and other superior vision center damages that frequently lead to what is designated *profound blindness*. The hope in these cases is to combine the increasing knowledge we have about the biology and the anatomy of our vision system with the amazing advances in science and technology, to set up a new field that can be designated as *BioElectronic Vision*. This book is about this emergent field, with the final goal of achieving components that substitute for these damaged vision centers. The remainder of this chapter identifies the main causes of blindness, discusses the primary concepts of *Bioelectronic Vision*, and references the research efforts around the world to design visual neuroprostheses.

1.1 Main Causes of Blindness

The impairments of profound blindness may have origins in degenerative retinal diseases or in brain injuries that affect the superior vision centers due to accidents or to direct surgical intervention (e.g. for a tumor removal). Figure 1.1 shows the distribution of the principal causes of blindness in the world, and their prevalence based on data collected from the World Health Organization, authority for health within the United Nations Organization, in 2002 [World Health Organization (2004)].

One of the major causes of vision impairments in the world is cataract. Cataract is normally related to the aging process and is characterized by opacity of the eye's lens, which impedes the regular flow of light. The actual treatment for this disease consists of a surgical intervention to replace the opaque lens with an artificial intraocular lens.

The second major cause of blindness is glaucoma. Glaucoma occurs when the aqueous humor does not drain out correctly and the pressure within the eye becomes too high, compromising the blood vessels of the optic nerve's head, and eventually the axons of the ganglion cells, which results in the death of these vital cells. The reduction of intraocular pressure is imperative to avoid total blindness. This disease affects the retinal nervous system and can cause permanent damage.

The third cause of blindness worldwide is the age-related macular degeneration (AMD). In some persons, the macula, which is responsible for

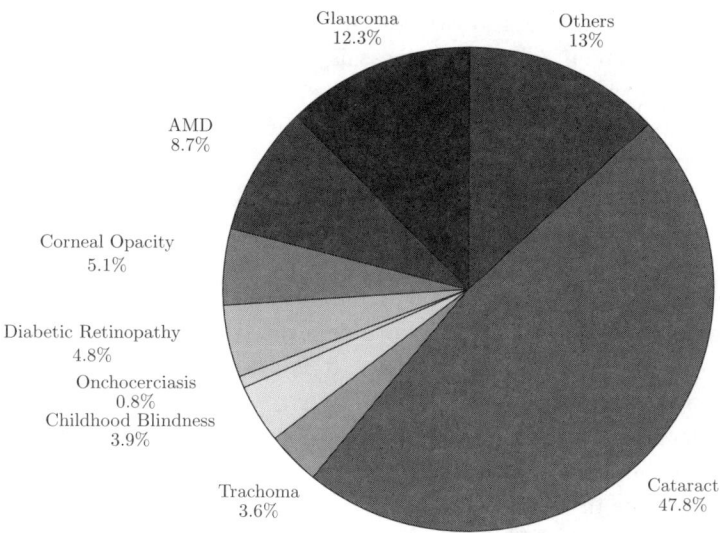

Fig. 1.1 Causes of blindness in the world in 2002 (data from [World Health Organization (2004)]).

the perception of fine detail in the center of the visual area, degrades with age for unknown reasons. The pigment epithelium behind the retina degenerates, forming drusen, and leaks fluid behind the foveal macular area. The cones in the fovea then die, which causes central vision loss and makes it impossible to read or see fine details. The AMD is a major cause of blindness in developed countries due to the high number of people above 70 years of age.

The graphic in Fig. 1.1 shows that the next cause of blindness is corneal opacity, which occurs when the cornea becomes scarred, preventing light from passing through the cornea to the retina, and causing, in some cases, the cornea to appear white or clouded. Corneal opacity can be caused by infections like conjunctivitis, or by the herpes virus, measles, injury, or inflammation of the eye caused by a stroke or a chemical agent. In many cases, it can be reversed by adequate treatment, which may include surgery.

Trachoma, another cause of blindness, is an infection caused by an organism (*Chlamydia trachomatis*) that can be treated with antibiotics. It is a common cause of blindness worldwide, but rare in developed countries.

A significant percentage of blindness is caused by diabetes, which is a serious problem in industrialized countries. Approximately 90% of all diabetic patients have retinopathy after twenty years. Diabetic retinopathy is characterized by anomalies in the blood vessels that get blocked and leak, or multiply in an uncontrolled manner, leading to irreversible blindness.

A major cause of blindness among children is the deficiency of vitamin A, particularly in children under 5 years. Included in this percentage is blindness caused by premature birth, infant retinopathy and cataracts. Blindness among children is a major problem due the length of time they will have to contend with the disability. It is estimated that 1.4 million children below age 15 are blind.

Onchocerciasis is responsible for blindness particularly in African and Latin America countries. Onchocerciasis is a disease transmitted by a parasite spread by flies in riverside areas.

The remaining causes of blindness are grouped in the general class "others". This includes a terrible disease called retinitis pigmentosa (RP), which presently has no cure. RP is an inherited disease that causes degeneration of the retina and pigment excess. First, it provokes night blindness, then tunnel vision and, as more of the peripheral retina becomes damaged and the rods die, progresses gradually to total blindness.

The blindness distribution is not geographically uniform. About 90% of visually impaired people live in developing countries. Statistics suggest that females have a higher risk of being visually impaired. In terms of age, it is estimated that about 82% of visually impaired people are more than 50 years old. "Vision 2020: The Right to Sight" is a global initiative for the elimination of avoidable blindness, launched jointly by the United Nations World Health Organization (WHO), the International Agency for the Prevention of Blindness (IAPB) and international eye care institutions and corporations. One of the largest and most productive eye care facilities in the world is the Aravind Eye Care System, which was established in 1976 in Madurai, India. It has treated over 2.3 million outpatients and performed over 270,0000 surgeries. They mainly serve people living in rural India, and they were the recipient of the first edition of the António Champalimaud [1]

---

[1] The Champalimaud Foundation (http://www.fchampalimaud.org), which is based in Lisbon, was created in 2004 at the bequest of the late Portuguese industrialist and entrepreneur António Champalimaud, to support individual researchers and research

Vision Award in 2007.

At the same time, several projects involving multidisciplinary research groups have been promoted to develop and demonstrate the feasibility of artificial vision systems. There are a handful of initiatives around the world devoting significant resources and attention to the research and development of visual neuroprostheses. The huge challenge of artificially restoring vision to the blind poses several engineering and biological problems that are difficult to overcome and also requires clinical human testing. The challenge is being addressed primarily by those in academia, but the effort comes from all over the world, as it can be seen in Fig. 1.2. Due to the initial cost of such prostheses, people who live in industrialized countries are expected to be the initial beneficiaries of this research. Thus, blind people affected by diabetes (AMD) and RP could be the first to take advantage of those neuroprostheses.

1.2 Main Components of a Bioelectronic Vision System

As will be seen later in this chapter and in Chap. 6, bioelectronic vision systems are supported on two main classes of visual neuroprostheses: *i)* retina neuroprostheses are suitable when the front end of the retina is functioning properly, while *ii)* a cortical neuroprosthesis is the last hope when the retina, including the optic nerve, is not functional at all, and only the brain vision centers remain (see Fig. 1.3). In this latter case, the neuroprosthesis directly interfaces with the visual processing center in the brain, the area (V1) of the visual cortex. In the profoundly blind, the optical signaling pathways are irreversibly damaged. Thus, the neuroprosthesis is a substitute for these patients' entire vision system. The concept and components of a bioelectronic vision system supported on a complete visual neuroprosthesis that directly interfaces with the brain is depicted in Fig. 1.3 and in Fig. 1.4.

The first component of the visual system is the eye. The eye is responsible for transducing light into neural signals, consisting of electrical impulses, that are then transmitted to the brain for further information extraction. Roughly speaking, the eye is composed of an optical system that focuses light on the retina, a neuronal tissue. Light patterns are encoded into electrical signals and the neuronal processing starts in the retina, as the retina is, in fact, an extension of the brain. A visual neuropros-

---
teams working in medical science, and particularly in the field of neuroscience.

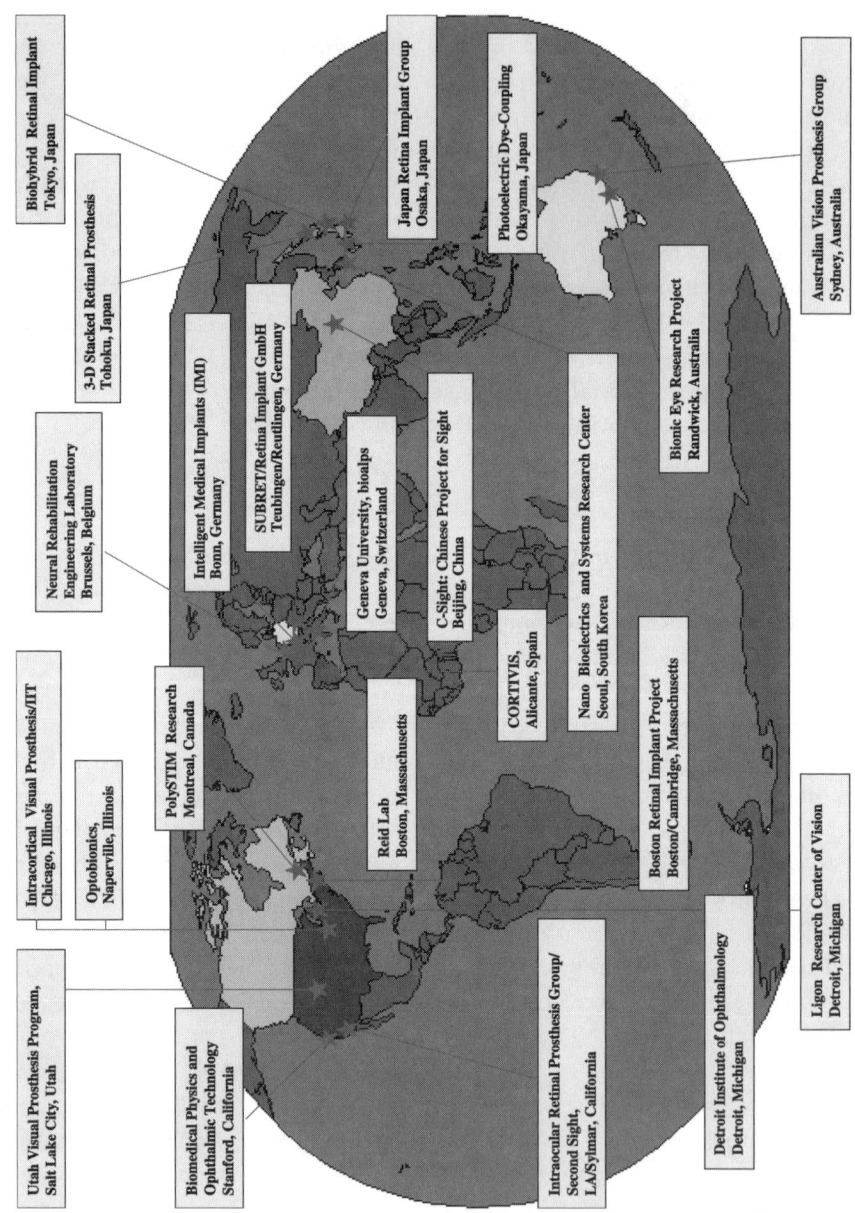

Fig. 1.2 World map of vision prosthesis groups (from [Hessburg and Rizzo III (2007)]).

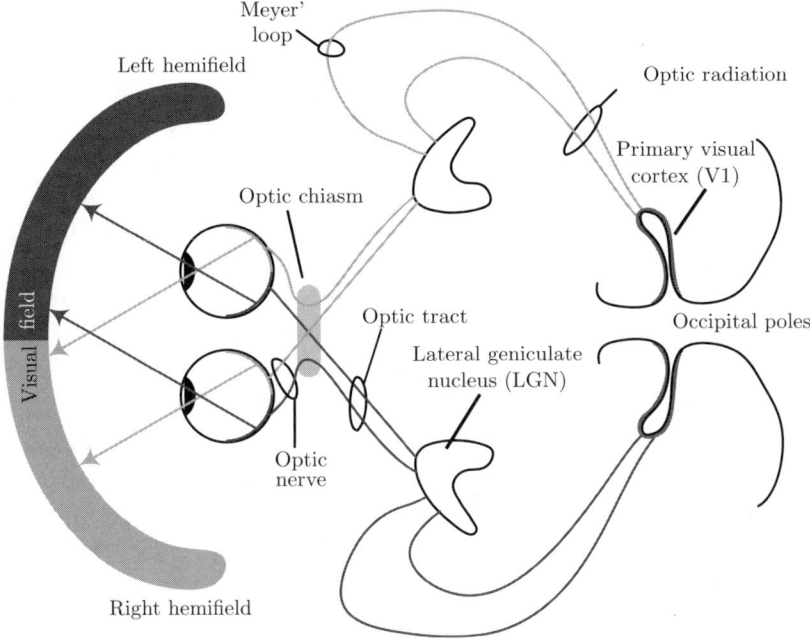

Fig. 1.3 Diagram of the human visual system.

thesis must model the optical eye system, which currently does not pose major technological difficulties, and the neuronal processing occurring in the retina, which is where the true challenge lies. Therefore, a significant step in the development of any visual prosthesis is the selection and evaluation of the adopted retina model. Two different approaches have been proposed for these models: the structural models try to mimic the biological systems based on knowledge about its physiological composition and operation, while the functional models attempt to replicate the functions performed by the retina but are not motivated by the characteristics of the biological systems themselves [Wulf (2001)]. Images in the visual field are mapped according to these models into a set of discrete signals that are then used to stimulate the visual cortex in the brain.

As can be seen in Fig. 1.4, a bioelectronic vision system includes a set of components which, depending on the class of visual neuroprosthesis, can be biological structures or their electronic circuit counterparts. For example, an external video acquisition device is usually required for capturing the visual field image and converting the light patterns into electrical signals. However, when an array of stimulation electrodes is placed in the subretinal space, the image falling on the retina and its light impulses are converted into electrical currents by microphotodiodes that directly replace and function in the place of the damaged photoreceptors.

The digital signal processing system transforms the visual space image into a set of discrete signals, according to the retina model and taking into account the visuotopic organization of the target structure (retina or cerebral cortex). A module to transmit power and control signals to the implanted electronics is usually required. This module provides energy and controls the stimulator that interfaces with the nervous system to induce the perception of phosphenes, which is an entoptic phenomenon characterized by the sensation of seeing light. The implemented electronics, based on integrated circuits and (micro)electrodes, will ultimately replace the function of their biological counterpart elements. Let us zoom in on the components represented in Fig. 1.4.

Image acquisition is currently a common task in engineering. For cortical or optic nerve neuroprostheses, a general, small and fully functional digital camera is well-suited for a visual prosthesis in terms of dynamic range, sensitivity and depth of field and it is aesthetically pleasing. In a retinal neuroprosthesis, the image encoder can be integrated into the neural interface, and can reside at the plane of the retina, with the advantage that the eye optics are used to project the image onto the encoder.

In the signal processing block, the biggest challenge is the visuotopic mapping of the visual space onto the target visual structure, particularly the visual cortex. This is a somewhat complicated task due to the uniqueness of this map among individuals and because it is conformal only at a low resolution; for high spatial resolutions, this mapping seems to be locally random. Therefore, parameterizable models have to be developed for implementing this module and properly stimulating individuals. This is a somewhat more complicated task due to the plasticity of the visual pathways and the different possible combinations between electrodes and phosphenes elicited. Based on the developed models, the electronics of this

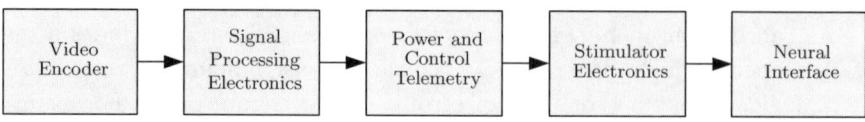

Fig. 1.4 Main components of a visual neuroprosthesis.

module transform the image into a discrete set of signals that drive the stimulators. To adapt the intensity of the incoming light signals into the range level of the neurons being stimulated, an automatic gain control (AGC) can function similarly to the photoreceptors. The first two components of the visual neuroprosthesis can be combined into a single device attached to a set of eyeglasses, while the elements of the visual neuroprosthesis described in the following are likely to be located inside the patient.

The information from the visual scene must be conducted to the implant, but a retinal neuroprosthesis and a cortical neuroprosthesis accomplish this via different methods. There are two main ways to transmit signals through the skin: percutaneous connectors [Dobelle (2000)] or using radio frequency (RF) telemetry [Piedade et al. (2005)]. On one hand, the percutaneous connectors have the advantage of being more robust and obviate the use of multiplexers, but on the other hand, they have the drawback of being a source of infections. A radio frequency link also has the challenge of communicating both power and control signals and the requirement that it must be bidirectional. In a retinal neuroprosthesis, a percutaneous connection would have to pass to the outside of the eye through the sclera. For the case of a retinal neuroprosthesis, a laser can also be used to transmit power and information to the implanted circuits used to stimulate the target cells [Weiland et al. (2005)]. All types of connections have a series of constraints such as bandwidth, which increases with the number of electrodes in the neural interface, and the transmitted power must also be limited to avoid causing damage to the tissues by heating.

The next component in the chain of Fig. 1.4 is the neural stimulator that must be capable of exciting multiple electrodes at the same time in order to evoke consistent phosphenes. It receives power and data through a telemetry sub-system, and must be capable of controlling the amount of power delivered to avoid damaging neighboring tissues. Furthermore, it should be able to circumvent malfunctioning electrodes. The implementation of this module on chips involves a trade-off between the processing capabilities and power consumption; increasing the process capabilities of the chip diminishes the required link bandwidth, but increases power consumption and potential failures [Maynard (2001)]. The most appropriate technology today, based on size and power consumption, is a digital VLSI circuit [Warren and Normann (2003)].

The last element in a vision prosthesis is the interface with the nervous system. The neural interface establishes the bridge between the nervous system and the external electronics. It mediates the transduction between the electrical currents generated by the electronic device into ionic currents that flow inside the human body. For the retinal neuroprosthesis, the neural interface options range from silicon chips to specific developed ceramic materials [Wu (2006)]. For cortical interfaces, oxidized iridium is a candidate material because it has shown good biocompatibility and acts as a good electronic to ionic current transducer. There are also other compatibility issues related to the neural interface that must be taken into account [Warren and Normann (2003)].

In conclusion, before reaching the ganglion cell layer in the retina or the visual cortex, the visual signal has already been subjected to a series of processing stages. When the interface with the visual stimulus is made at the level of the ganglion cell layer, as in the case of an epiretinal neuroprosthesis, the output information must be identical to the output produced by a healthy retina; the transformation of information about the visual space to the retinotopic space is done by modeling the neural processing of the retina. For the case of a cortical neuroprosthesis, the signal processing occurring along the visual pathway should take place such that an adequate stimulus for the neural interface is generated.

1.3 Classification of Visual Prostheses

The effort to provide the profoundly blind with some kind of vision has led to some results [Rizzo III and Wyatt (1997)]. Throughout the world, several research groups and consortia dedicate their efforts to designing vision prosthesis capable of conveying to the blind "some kind" of vision. The words "some kind" are used frequently by scientists when referring to this goal and reflect the huge extension of this task framed by the complexity of the human visual system, ranging from the retina's neural circuitry to the deep brain processes involved.

Some unconventional approaches and electronic devices have been proposed to convey vision to visually impaired people. In some of these devices, the visual information is converted to auditory [Arno et al. (1999)] or tactile signals and is afterwards communicated to the brain. One somewhat curious device is one that consists of a flexible cable with a matrix of electrodes at the end that is placed against the patient's tongue, and a pattern of electrical impulses stimulates its sensitive nerves [Weiss (2001)].

Bioelectronic visual systems are supported on visual neuroprostheses that interface with the following neural structures: *i*) the photoreceptor layer of the retina; *ii)* the ganglion cell layer of the retina; *iii)* the optic nerve, and *iv)* the visual cortex [Warren and Normann (2003)]. Thus, there are three types of prostheses that still uses some part of the human visual system: retinal and optic nerve neuroprostheses at the eye level, and cortical neuroprostheses at the brain level. The retinal neuroprostheses use the remaining functioning parts of the retina to send the visual signals to the brain, the optic nerve neuroprostheses stimulate what is left of the optic nerve, and the objective of the cortical neuroprostheses is to inject the visual signals directly into the visual cortex.

The type of approach used in a visual prosthesis is related to the type of blindness. In one extreme, the blindness may be caused by damage to the superior retina layers as a consequence of the early stages of diseases like RP. Such blindness may also stem from some kinds of macular degeneration, where the principal injuries occur at the photoreceptor layer, but the ganglion cell layer remains mostly intact, which allows for the (re)usage of these cells. In this type of blindness, a retinal neuroprosthesis can be used. At the other extreme, there is what is called *profound blindness*, where the ganglion cell layers and the optic nerve are irreversibly injured and incapable of transmitting any kind of nervous signals. In this case, the re-establishment of some sort of vision can only be done by the circumvention of the optic nerve, and the remaining option is to stimulate the visual cortex directly with an electronic device. This is where the cortical neuroprostheses come into play.

An intermediate situation is the utilization of the optic nerve to conduct the visual signals to the brain. Some of the retinal diseases leave a significant number of intact ganglion cells dispersed along the retina, whose axons converge to form the optic nerve. The strategy here is to induce the visual signals in this great bunch of ganglion cells at the optic nerve, expecting conductive axons to be excited by the stimulus.

In the remainder of this chapter, we present the main characteristics of the prostheses that interface with the retina and directly with the visual cortex. The signal processing modules associated with these neuroprostheses will be discussed in the next chapters. All this information will be put together in Chap. 6, where the design and the implementation of bioelectronic vision systems will be discussed.

1.3.1 Retinal Neuroprosthesis

There are the two kinds of retinal implants, depicted in Fig. 1.5. In a *subretinal* implant, the prosthesis is placed between the pigment epithelial layer and the outer layer of the retina, which contains the photoreceptors cells. In contrast, the epiretinal device is placed directly against the ganglion cells and their axon layer, bypassing the rods and cones, and directly stimulating the inner retina; this is a more invasive technique.

Unlike the subretinal implant, the *epiretinal* implant does not use any remaining network of the retina for information processing. Thus, the epiretinal sensor has to encode visual information as trains of electrical impulses that are then conveyed by an electrode array directly into the ganglion cell axons, which converge to form the optic nerve. This spatiotemporal stimulation pattern of electrical impulses must represent the visual information in such a way that it can be understood by the brain's visual cortex. On the other hand, the information-transfer characteristics of the epiretinal implant are more amenable to external control, while the subretinal implant requires intact original optics [Zrenner (2002)]. A relevant example of the development and testing of a subretinal implant is reported in [Chow et al. (2004)], and an example of an epiretinal implant can be found in [Rizzo III and Wyatt (1997); Humayun et al. (1999)].

Nevertheless, retinal neuroprostheses require the presence of viable cells in the inner retina. Therefore, diseases limited primarily to the outer retina are potentially treatable with a retinal neuroprosthesis. Margalit et al. references [Margalit et al. (2002)] and Weilandet al. [Weiland et al. (2005)] present an extended overview of retinal neuroprostheses.

One of these prostheses, entitled the "Bionic Eye", uses a new ceramic material to replace the retina's photoreceptors, which acts as an optic detector that transduces light into electrical impulses by means of the photo-ferroelectric effect [Wu (2006)]. This material is directly implanted in the patient's eye and, under optic illumination, generates a photo-current that directly excites the retinal ganglion cells. It seems to be bio-compatible, and it can be used in retinal dystrophies, where the optic nerve and retinal ganglia are intact. For example, for RP, the ceramic material is used to directly stimulate the retinal ganglia.

Another example of a retinal neuroprosthesis is the artificial silicon retina (ASR) microchip [Chow et al. (2004)], which uses well-known silicon technology [Optobionics Corporation (2006)]. The ASR microchip is a silicon-based device with a diameter of 2 mm that contains approximately

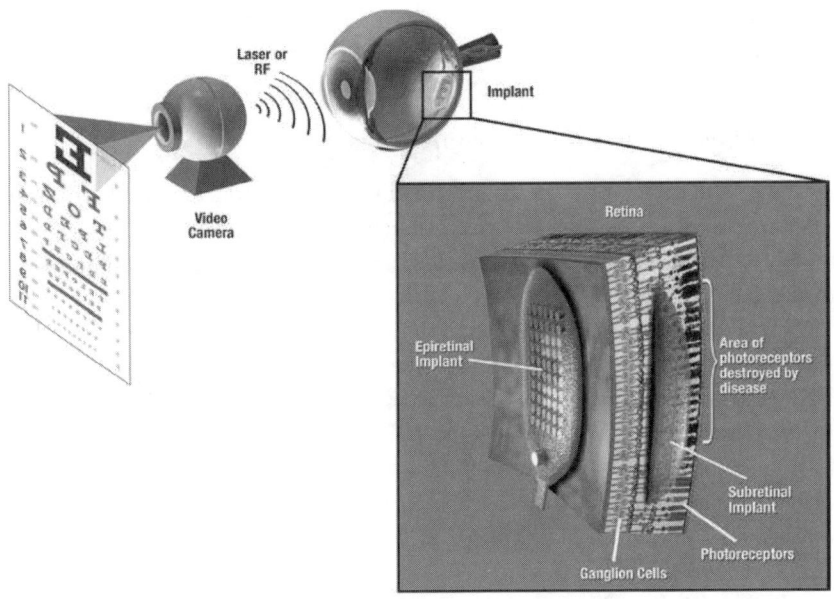

Fig. 1.5 Main components of epiretinal, retinal and subretinal prostheses. (From [Weiland et al. (2005)])

5000 microelectrode-tipped microphotodiodes and is powered by incident light. The ASR microchip was implanted in the eyes of patients with retinitis pigmentosa. Patients did not show signs of implant rejection or infection, and all of them demonstrated improved visual function.

There are other examples of subretinal implants. For example, one consists of a chip ($3 \times 3 \times 0.1$ mm, 1500 microphotodiodes, amplifiers and electrodes of 50×50 μm, spaced 70 μm) and a 4×4 array of identical electrodes, spaced 280 μm apart, for direct stimulation. These were chronically implanted next to the foveal rim of two blind retinitis pigmentosa patients [Zrenner et al. (2006)]. The implant was removed in one patient after 4 weeks, but the other patient decided to keep the implant. Patients reported small, yellowish or greyish phosphenes for individual electrode stimulation and they were able to differentiate spatial patterns such as lines, angles and bright squares.

There is a second type of approach at the level of the visual system whose functional principle is to directly stimulate the optic nerve, circumventing the retinal layers entirely, which may be damaged. These devices are implanted around the optic nerve to stimulate the fibers electrically [Brélen et al. (2005)]. This type of prosthesis also requires the development of a retina model.

1.3.2 Cortical Visual Neuroprosthesis

Cortical visual neuroprostheses are bioelectronic systems that use the visual cortex in the brain as the interface between the electronics components and the biological visual pathway. A last resource for blind individuals who cannot benefit from a retinal neuroprosthesis is the direct stimulation of the visual cortex. This is the last hope when the retina is not functioning at all (including optic nerve failure) because retinal neuroprostheses rely on the optic nerve to transmit electrical signals from the eye to the visual cortex. The visual cortex is a brain vision processing center, and it is well positioned for direct stimulation. This kind of neuroprosthesis includes all the electronic components presented in Fig. 1.4 to substitute the biological counterparts, shown in Fig. 1.3.

The first permanent device developed and applied for chronic stimulation of neural tissue was accomplished in 1968 [Brindley and Lewin (1968)]. The device had 80 electrodes, each with its own controlling unit (receiver). Using this system, the feasibility of a permanent cortical vision neuroprosthesis was demonstrated as the electrical stimulation of the occipital lobe of the human cortex caused subjects to perceive phosphenes.

Despite this initial success [Dobelle and Mladejovsky (1974)], the surface electrodes have a number of drawbacks. The electrical current necessary to induce a phosphene is relatively high (on the order of a milliampere); consequently, the distance between the electrodes must be considerable in order to minimize their interactions due to current spread. Unfortunately this degrades the spatial resolution. Moreover, current injections can produce short term and long term complications depending upon the levels of the currents that are injected [Agnew and McCreery (1990)].

Two main groups worked during the 1990's towards a cortical vision prosthesis. One was based at the National Institute of Health (NIH) in Washington, D.C., and the other at the John Moran Laboratories in Applied Vision and Neural Sciences at the University of Utah. Both groups tried to overcome the problems mentioned above by employing penetrating microelectrodes instead of using surface electrodes on the visual cortex. One example of such a microelectrode array was manufactured using semiconductor technology [Maynard (2001)] and was developed at University of Utah; it is known as the Utah Electrode Array: 10×10 microelectrodes, each 1.0 − 1.5 mm long, dispersed in a square grid contained in a package with dimensions 4.2 × 4.2 mm [Maynard et al. (1997); Normann et al. (1999)]. The silicon micromachining and micromanufacturing technologies allow the fabrication of small arrays with a large number of microelectrodes capable of stimulating only the neurons nearest to the electrode with a small amount of current (on the order of dozens of μA). The major concerns with the microstimulation are related to the biocompatibility and long term functioning of the inserted microelectrode array.

Research is ongoing to design and develop cortical visual neuroprostheses through intracortical stimulation, but none of these prostheses has been permanently applied for chronic stimulation. A European research project, CORTIVIS, has been conducted over the last few years to design and develop a complete visual neuroprosthesis designed to restore useful vision to profoundly blind people [Project CORTIVIS (2006)]. It performs intracortical microstimulation through one or more Utah Electrode Arrays implanted into the primary visual cortex. The system is composed of a primary unit located outside the body and a secondary unit, implanted inside the body, that communicate with each other using wireless communication technology. A prototype of the proposed system is presented in detail in Chap. 6.

1.4 Conclusions and Further Reading

In 2002, it was estimated that more than 161 million people were visually impaired, of whom 124 million people had low vision and 37 million were blind. However, refractive error as a cause of visual impairment was not included in these figures, which implies that the actual global magnitude of visual impairment is greater [World Health Organization (2004)].

Figure 1.1 shows the distribution of the principal causes of blindness in the world and their prevalence based on data collected from the World Health Organization, the authority for health information within the United Nations Organization, in 2002 [World Health Organization (2004)].

Retinal neuroprostheses require the presence of viable cells in the inner retina. Therefore, diseases restricted primarily to the outer retina are potentially treatable with a retinal neuroprosthesis. Margalit et al. [Margalit et al. (2002)] and Weiland et al. [Weiland et al. (2005)] present an extended overview of retinal neuroprostheses. One retinal neuroprosthesis example, called the "Bionic Eye", uses a new ceramic material to replace the retina's photoreceptors, acting as an optical detector, transducing light into electrical impulses via the photo-ferroelectric effect [Wu (2006)]. Another example of a retinal neuroprosthesis is the artificial silicon retina (ASR) microchip [Chow et al. (2004)], which uses the well-known and dominant silicon technology [Optobionics Corporation (2006)].

There are other examples of subretinal implants, namely the one that consists of a chip, 16 direct stimulation (DS) electrodes, and a power line implanted subdermally, connecting the chip to an external energy supply [Zrenner et al. (2006)].

The effort to provide some kind of vision to the profoundly blind already has some history [Rizzo III and Wyatt (1997)]. The first permanent device developed and applied for chronic stimulation of neural tissue was accomplished in 1968 [Brindley and Lewin (1968)]. Although it was observed that the electrical stimulation of the occipital lobe of the human cortex causes a subject to perceive phosphenes, it was also concluded that current injection can produce short term and long term complications [Agnew and McCreery (1990)].

Some of the experiments conducted at the Dobelle Institute, when cortical neuroprostheses were initially proposed in 1974 [Dobelle and Mladejovsky (1974)], involved implantation of prototypes in blind people [Dobelle (2006)] and demonstrated that focal epileptic activity can be induced by electrical stimulation. Therefore, penetrating microelectrode arrays manufactured using semiconductor technology [Maynard (2001)], such as the microelectrode array, were developed at the University of Utah, and are now known as Utah Electrode Arrays. These microelectrodes are deeply inserted into the virtual cortex, using a pneumatic insertion technique, to allow for intracortical stimulation with low currents and without provoking major injuries [Maynard et al. (1997); Normann et al. (1999)]. The silicon micromachining and micromanufacturing technologies allow the fabrication of small arrays with a large number of microelectrodes capable of stimulating only the neurons nearest to the electrode with a small amount of current (on the order of dozens of micro amperes). Donoghue [Donoghue (2002)] provides a general perspective of cortex electronic interfaces.

A European research project, CORTIVIS, has been conducted in the last few years to develop a visual cortical neuroprosthesis based on the Utah Electrode Array [Project CORTIVIS (2006)]. A prototype of the proposed system has been developed. More detailed technical information can be found in Chap. 6 and in [Piedade et al. (2005)].

Table 1.1 summarizes the main pros and cons of the different approaches described in this chapter to develop visual neuroprostheses. In the next chapter we will present the main features of the human visual system which are important to introduce, in Chap. 3 and in Chap. 4, the signal processing tools and the retina models adopted to design bioelectronic vision systems.

Exercises

1.1. Identify the main components of the human visual system represented in Fig. 1.3 and enumerate their main functions and characteristics.

1.2. Based on the 2002 world population of 6 thousand million, predict the relative percentages of people who were visually impaired and who were blind at that time.

1.3. Enumerate the main classes of visual neuroprostheses referred to in the text.

1.4. By taking as a reference the main classes of visual neuroprostheses presented in the text and enumerated in the previous section,

- **1.4.1** associate the different classes of neuroprostheses with the types of vision impairments they would be used to treat;
- **1.4.2** identify to which class of visual neuroprostheses the "Bionic Eye" and the ASR microchip belong;
- **1.4.3** identify the class of visual neuroprosthesis targeted in the CORTIVIS project.

1.5. List the main components of a bioelectronic vision system and the main electronic components associated with visual neuroprostheses. Comment on the main challenges for the implementation of these prostheses.

1.6. Elaborate on the advantages and disadvantages of the different approaches to develop visual neuroprostheses, starting with the information provided in Table 1.1.

1.7. Identify the main advantages of the penetrating microelectrodes, such as the Utah Microelectrode Array, compared to the surface electrodes used, for example, by Dobelle.

1.8. Enumerate the two main approaches used to develop retina models, referring to their main differences.

1.9. Choose three research groups on the world map shown in Fig. 1.2, *from different continents, participating in this kind of research. Consult the internet and describe the research they have performed and their achievements in the field of visual neuroprosthesis development.*

Table 1.1 Main pros and cons of visual prostheses approaches.

Advantages	Disadvantages
Visual Cortex	
Only approach for non-functional retinas and/or optic nerves	Stimulation far from photoreceptors
Implant site robust and protected by skull	Possibly poor visuotopic organization
Easy surgical access	Multiple feature representations in V1 (color, lines, motion, ocular dominance)
High density electrode implantation	
Phosphene thresholds are low (1-10 μA)	Societal phobias about "brain implant"
	Consequences of surgical complications
Epiretinal	
Stimulating close to photoreceptors: uses native processing in thalamus and cortex	Requires functional optic nerve pathway
Fewer surgical complications than cortical implants	May stimulate optic nerve fibers – greatly complicates visuotopic organization
Saccadic eye motions cause sheer loads on implanted arrays	
Difficult to adhere electrode array to retina	
Subretinal	
Stimulating closest to photoreceptors – uses retinal, thalamic and cortical processing	Requires functional retina and optic nerve pathway
If bipolar cells can be directly stimulated, retinotopic organization should be preserved	Blockage of nutrients from choroid by the implant
	Very complex surgical access
Fewer surgical complications than cortical implants	Cannot stimulate cells passively with microimplants (requires external power)
Optic Nerve	
Fewer surgical complications than cortical implants	Requires functional optic nerve pathway
	Visuotopic organization requires placing electrodes at many closely spaced regions of the optic nerve
	Complex electrode array to provide patterned vision
	Very difficult surgical access

Chapter 2

The Human Visual System

2.1 Introduction

The visual system of vertebrates is among the most complex that exists in nature. The visual system is composed of the eye, whose principal function is to gather a pattern of light that assembles an image from the surrounding world and to convert this image into neural signals, and the brain visual centers that process the neural signals and extract the intended visual information. Since the visual system in humans, and in mammals in general, is very complex, a substantial amount of research effort in the last decades has been directed at understanding the various aspects of the system, ranging from the physical to the biological and psychological processes involved.

After a short description of the eye composition, this chapter delves into a somewhat detailed description of the anatomy and physiology of the human retina. It serves as a basis for understanding the different processes occurring in the retina to provide a functional evaluation of the retina models and to gain insight into the challenges encountered in modeling such a complex and intricate network of neurons. It also helps to devise methods and simplifications that can be applied in a computational retina model. Then, a summarized description of the visual pathway is also provided with a presentation of the principal visual processing centers, ending in the visual cortex.

Finally, some relevant issues related to retinal modeling are discussed. In addition, a brief overview of the basic processing blocks commonly used, as well as the taxonomy usually employed to classify those models is described. However, as the retina is a neural circuit, we start with a general description of the neuron anatomy and dynamics.

2.2 The Neuron

To have a better insight into the vision system, and to understand several phenomena in the retinal neural cells' responses, it is important to start with a general overview of the neuron anatomy and the mechanisms involved in information encoding and communication.

2.2.1 Neuron Anatomy

The neuron is the basic unit of information processing and is the building block of neural circuits, such as the retina and brain visual centers. The neuron is made up of three basic components, as shown in Fig. 2.1: the cell body, or *soma*; an extension, called the *axon*; and the *dendrites*. Dendrites look like the branches of a tree; they receive messages from other cells and communicate it to the soma. A single neuron can have more than 2000 dendritic branches, establishing connections between tens of thousands of other cells. Within the cell body (soma) is the nucleus, which contains the genetic material. The main role of the soma is to process all the information collected by the dendrites; if the sum of the electrical signals collected by the dendrites is strong enough, the neuron will fire an action potential. The axon transmits messages from the cell body to other neurons. The axon looks like a long tail, and can extend farther than 1 m (the largest axon in humans runs from the base of the spine to the big toe of the foot), and can be as wide as 1 mm. At its end, the axon divides into fine branches – the *axonal terminals* or *pre-synaptic terminals* – that make contacts with neighboring neurons. In many neurons, portions of the axon are covered by a myelin sheath. Myelin is a fatty substance whose role is to increase the speed and strength of the signal that travels down the axon, in addition to protecting the axon from external assaults.

The point of contact between two neurons is called the synapse (see detail in Fig. 2.1). A synapse is composed of a narrow space, called the *synaptic cleft*, between the axon ramification ends of the neuron that transmit a signal, the *pre-synaptic neuron*, and the dendrite of another neuron – the *post-synaptic neuron*. Most pre-synaptic terminals end on the dendrites of the post-synaptic neuron (axodendritic synapse), but the terminals can also target the cell body (axosomatic), and less frequently, the beginning or end of the axon of the receiving cell (axoaxonic synapse) (see Fig. 2.1). Synapses can be also found between neurons and other cells, such as muscle cells and gland cells, with appropriate receptors. There are a number of different types of synapses found in the retina which can be classified into two main categories: chemical synapses and electrical synapses. Synapses which communicate through a *transmitter substance*, or *neurotransmitter*, are termed *chemical synapses*, and those in which two cells are electronically coupled together are termed *electrical synapses*. The specific sites where the neuron membranes connect in the electrical synapse are termed *gap junctions*.

A single neuron can receive synapses from a large number of other neural cells and its response can vary significantly, depending on which source cell (or set of cells) stimulated it. Moreover, the temporal history of stimulations can change the synaptic strength. This mechanism, typically referred to as *synaptic plasticity*, is an essential process that, in addition to roles in learning, is crucial for the physical building of the brain during its development and throughout an organism's lifetime [Dayan and Abbot (2001); Gerstner and Kistler (2002)].

Besides the nerve cells, or neurons, the nervous system is also composed of other types of cells, like glial cells that outnumber the neurons. Glial cells are substantially smaller and provide support and protection for neurons. They are active in surrounding neurons and holding them in place; in supplying nutrients and oxygen to neurons; in insulating one neuron from another; and in destroying pathogens and removing dead neurons. Glia also have important developmental roles, such as guiding migration of neurons in early development and producing molecules that modify the growth of axons and dendrites [Purves *et al.* (2007)].

2.2.2 Neuron Dynamics

When the integration of the signals gathered by the dendrites and delivered to the cell body surpasses a given threshold, an action potential is generated. An action potential is an electric pulse that is generated at the axon hillock, where the axon emerges from the cell body (see Fig. 2.1). This electrical signal travels along the axon to the axonal terminals and is refreshed along the way to prevent signal decay. This transmission and signal refreshing involves the movement of charged particles – namely ions – across the neuron membrane, and terminates with the release of a chemical substance into the synapse. The synaptically transmitted messages can be either excitatory or inhibitory.

There are several types of ions involved in the transmission of the action potential along a neuron's axon. The propagation of electric signals due

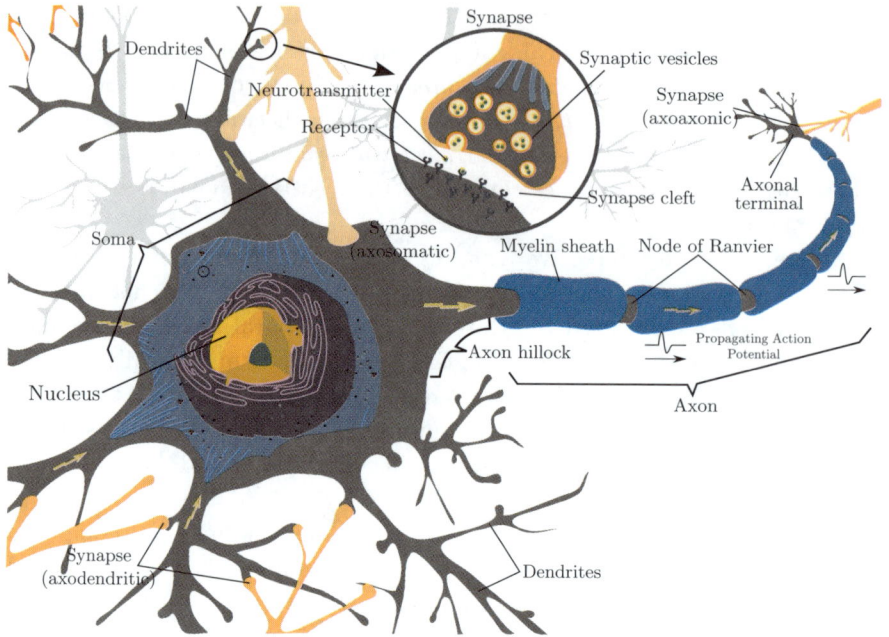

Fig. 2.1 Anatomy of a neuron.

to excitatory stimulation is achieved mainly by means of an exchange of sodium (Na^+) and potassium (K^+) (positive ions), between the inside and outside of the neuronal membrane. Other ions involved in the process are chloride (Cl^-) and organic ions (A^-), such as amino acids, proteins, and nucleotides, in smaller concentrations. Along the axon, the neuron possess ionic pumps that move the Na^+ ions to the outside of the cell (see Fig. 2.2(a)), while the K^+ ions are moved inside; however, a small outward leak of K^+ ions, larger than the inward leak of Na^+ ions, keeps the potential difference between the inside and the outside negative. It is thus said that the membrane is hyperpolarized. When an action potential is delivered by the soma, the Na^+ channel opens in the axon, as shown in Fig. 2.2(b), so that there is a influx of sodium ions into the cell; this raises the potential inside the cell membrane, making it depolarized. To compensate, the nearby K^+ ion channels open, leading to an efflux of K^+ ions (Fig. 2.2(c)), repolarizing the membrane. Eventually, both channels close (first the sodium, then the potassium) and the pumps re-establish normal conditions. All this is achieved in less than 2 ms. Furthermore, as

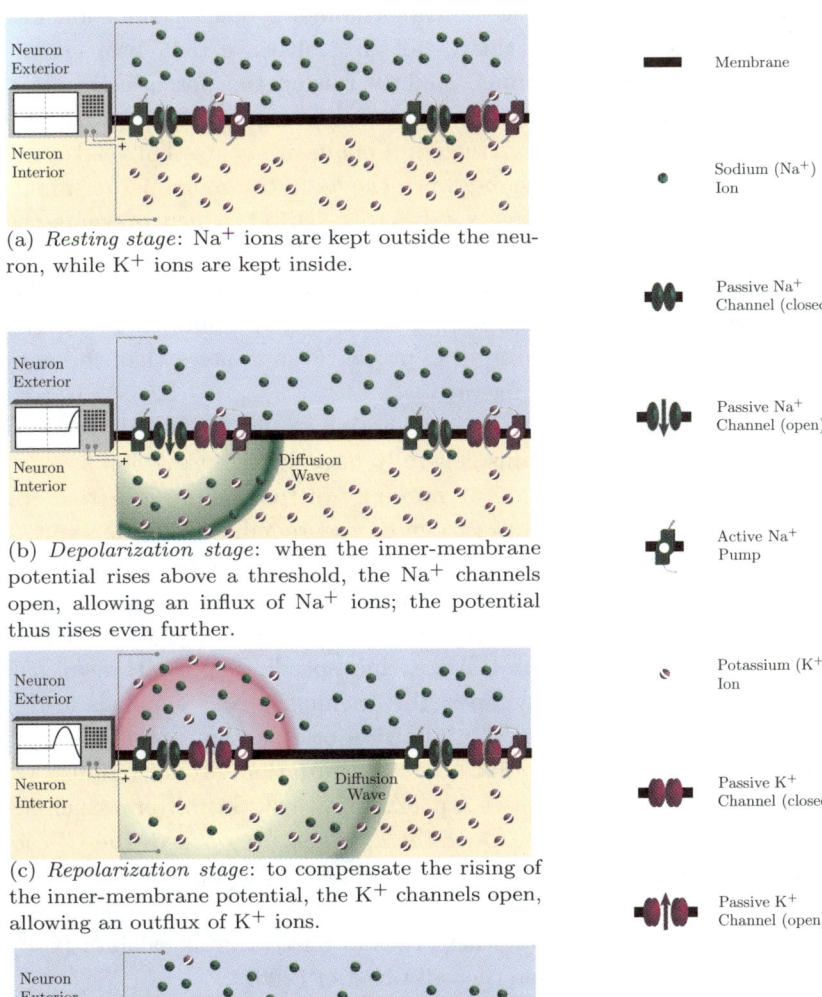

(a) *Resting stage*: Na$^+$ ions are kept outside the neuron, while K$^+$ ions are kept inside.

(b) *Depolarization stage*: when the inner-membrane potential rises above a threshold, the Na$^+$ channels open, allowing an influx of Na$^+$ ions; the potential thus rises even further.

(c) *Repolarization stage*: to compensate the rising of the inner-membrane potential, the K$^+$ channels open, allowing an outflux of K$^+$ ions.

(d) *Propagation of action potential*: the opening of Na$^+$ channels creates a depolarization wave that propagates to nearby channels; if it is strong enough, it will trigger the opening of the next channels and initiate a chain reaction.

Fig. 2.2 Propagation of an action potential along a neuron's axon.

illustrated in Fig. 2.2(b), once the Na^+ channel is opened, diffusion waves of Na^+ ions are generated; this depolarization wave then propagates to another nearby sodium channel, slightly raising the potential inside the cell. As a result, that nearby Na^+ channel opens, and the cycle repeats itself. This process therefore triggers a chain of events that leads to the propagation of the electric signal down the axon to the axonal terminals.

The ion channels also have a refractory period, which prevents them from re-opening in a short amount of time. This process is very important because it ensures that the cycle does not become astable. Without this feature, once the channels were activated, they would enter a cycle of opening and closing; moreover, this property guarantees that the action potential always propagates forward.

The speed of action potential propagation is usually directly related to the size of the axon. Big axons generally have fast transmission speeds as increasing the size of the axon allows more of the sodium ions that form the internal depolarization wave to enter and remain inside the axon. To overcome transmission speed limitations due to the limited size of the axons, myelin (mentioned above) is wrapped around many neurons' axons (see Fig. 2.1). The myelin sheath prevents the dissipation of the depolarization wave by inhibiting ion leakage, thus speeding up the transmission.

While in an excitatory synapse, the opening of Na^+ channels leads to the excitation of the cell and the initiation of an action potential, the inhibitory synapse works by preventing this initiation. To accomplish this, the pre-synaptic neuron releases a packet of neurotransmitters which activate chloride (Cl^-) channels. Once the channels are opened, the Cl^- ions flow into the neuron by diffusion; this lowers the potential inside the neuron and prevents the initiation of the action potential. The behavior of neurons is also influenced by other types of ions, such as calcium (Ca^{2+}) and magnesium (Mg^{2+}) [Gerstner and Kistler (2002)].

2.2.2.1 *Neural Communication through Chemical Synapses*

When the action potential reaches a chemical synapse at the axonal terminal, the pre-synaptic neuron communicates the information to the next neuron, the post-synaptic neuron. The process of information communication in the synapse is achieved as follows: *i)* the action potential triggers the opening of calcium channels, allowing for an influx of calcium (Ca^{2+}) ions into the pre-synaptic neuron; *ii)* the excess of Ca^{2+} ions causes the pre-synaptic neuron to release a packet of organic molecules, referred to as *chemical messengers* or *neurotransmitters*, into the *synapse cleft*, a small gap between the two neurons. These neurotransmitters are stored in synaptic vesicles (see Fig. 2.1), and serve as the output signal, translating the neuron's electrical signal into a chemical signal; *iii*) when released by the pre-synaptic neuron, the neurotransmitters traverse the synaptic cleft and bind to special proteins in the post-synaptic neuron that produce a local electrical signal called the *synaptic potential*. Unlike the action potential, the synaptic potential is not propagated; instead, it triggers the opening of special ion channels - sodium (Na^+) for excitatory synapses, or chloride (Cl^-) for inhibitory synapses; *iv*) these ions then enter the post-synaptic neuron, generating an electric signal. In an excitatory synapse, this signal propagates to the post-synaptic neuron body, eventually resulting in the initiation of another action potential; in an inhibitory synapse, the Cl^- ions work to prevent the generation of action potentials by inhibiting excitatory electronic signals. This synaptic communication can be mediated by different types of neurotransmitters [Gerstner and Kistler (2002)]. In the retina, the neurotransmitter passing through the vertical pathways, which run from photoreceptors to bipolar cells to ganglion cells, is glutamate, while the horizontal and amacrine cells send signals using various excitatory and inhibitory amino acids, including catecholamines, peptides and nitric oxide [Kolb (2003)].

The synaptic potential is not stereotyped, like the action potential; instead, its amplitude depends on the stimulus strength. The potential can be more positive than the neuron resting potential (depolarizing), enhancing the neuron's ability to fire an action potential (rendering them excitatory), or they can be more negative than the resting potential (hyperpolarizing), making the neuron less likely to fire an action potential (rendering them inhibitory).

2.2.2.2 Neural Communication through Electrical Synapses

In an electrical synapse the two neurons are electrically coupled at specific sites of the membrane surface – the *gap junctions*. A gap junction consists of several channels that allow ions, and small molecules, to pass from one cell to the other cell. Each channel consists of two subchannels, termed *connexons*, one from each cell. Each connexon is composed by six protein subunits termed connexin (see Fig. 2.3). Gap junctions can differ in the number of channels and in the types of connexons that compose them, however they always have a depolarizing effect.

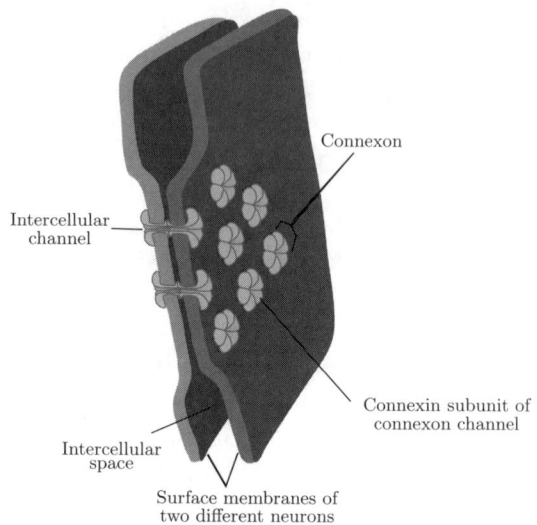

Fig. 2.3 Gap junction between two neurons.

Signaling by electrical synapses is very fast because the action potential pass directly from one neuron membrane to the other, through the gap junction, which allows the direct exchange of ions. The electric current is proportional to the potential difference between the presynaptic and the postsynaptic membranes.

2.3 The Human Visual System

The ancient Greeks thought that it was the light rays that emanate from the eyes and touch the objects, making them visible. This idea was related to the fact that, despite the fact that objects can be far away, they can still be sensed. It was only in the seventeenth century (1625) that the German Jesuit priest and physicist Christoph Scheiner (1573 – 1650) showed that it is the light entering the eye that produces the image. Since then, remarkable advances have been made in the knowledge and understanding of this marvelous sense, although there are still many unanswered questions. The sections that follow resumes what is currently known about eye anatomy and physiology.

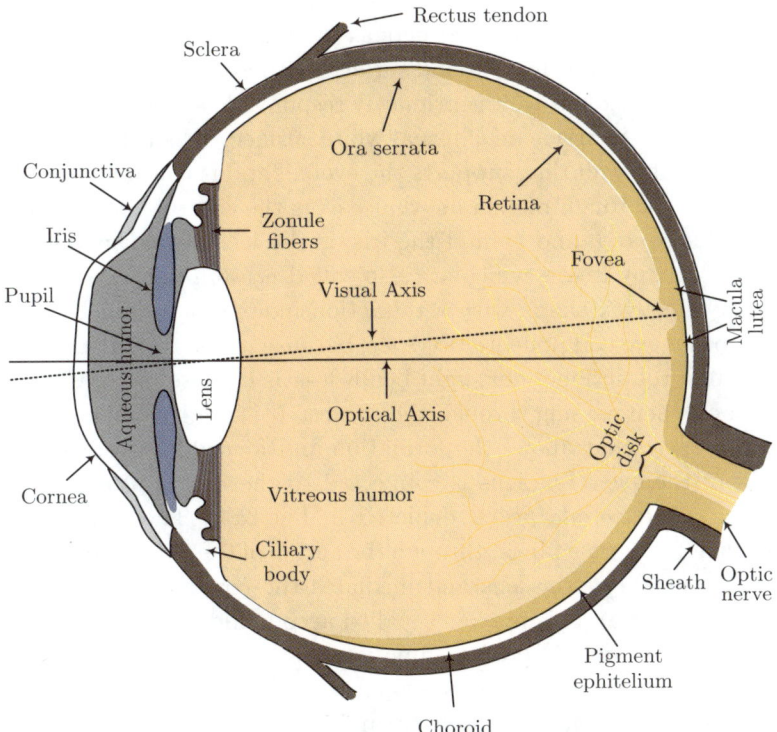

Fig. 2.4 Schematic section of the human eye.

2.3.1 *The Eye*

From a functional point of view, the eye is an optical system that gathers light and focuses it on its rear surface where the retina lies. Fig. 2.4 shows a schematic representation of the horizontal cross section of the human eye with the different parts labeled.

Looking into someone's eyes, it is possible to immediately identify several components. One is the pupil, an aperture in the center of the eye that appears dark due to the light absorbing pigments of the retina on the back. Around the pupil is the iris, a colored muscle that confers the color to the eyes, but whose color does not have any functional relevance. The iris muscle controls the size of the pupil according to the light conditions. The pupil diameter can range from 1.5 mm to 8 mm, becoming smaller in brighter light conditions.

Covering the external surface of the eye is the *cornea*. The cornea is a tough transparent membrane that acts as the first lens encountered by the light entering the eye. It has a round surface, with a refraction index of 1.38, and acts as a convex lens that is primarily responsible for bending light into the eye. Around the iris is an external white surface, called the *sclera*, that makes part of the wall that supports the eyeball and is continuous with the *dura mater* – the tough membrane that covers the central nervous system.

Partially covered, and behind the iris, is the lens, a transparent tissue consisting of many fibers, which are also called *crystalline*. The crystalline lens has a bi-convex shape with a refraction index of 1.4, which is higher than any other eye component. However, because it is surrounded by media with a similar refraction index, light bends less as it passes through the lens than it does when passing through the cornea. It is the lens, in conjunction with the cornea, that allows the formation and focus of a sharp image on the back of the eye. The lens is connected to the *ciliary body* by several ligaments, called *zonule fibers*, depicted in Fig. 2.4. The contraction or relaxation of the zonule fibers, through the action of the ciliary body muscle, changes the shape of the lens and mediates the constant focusing of the image on the retina. This process is called accommodation and constitutes the zonule fibers' most important function.

In the eyeball, there are three chambers of fluid. Between the cornea and the iris lies the *anterior chamber*, and between the iris and the zonule fibers lies the *posterior chamber*. These two chambers are filled with the *aqueous humour*. A third chamber, called the *vitreous chamber*, is located behind the lens and is filled with the *vitreous humour* that occupies the entire space between the lens and the retina, representing two-thirds of the eye's volume. The vitreous humour is a viscous fluid whose refraction index is identical to the eye's optical system, so that it does not bend light. The vitreous chamber is also responsible for the round shape of the eye.

Behind the vitreous humour is the retina, which covers 65% of the inside eyeball and is one of the most important components of the eye. This is where the image is focused and transduced to neural signals, to be posteriorly conducted to the brain by the optic nerve. The previously presented components of the eye have the unique function of focusing the image on the retina, playing a role similar to that of an optical system, while neural processing actually begins at the retina. Figure 2.4 displays the path followed by a light ray entering the eye's optics and hitting the retina, showing that the visual axis differs from the optical axis. The visual axis ends in a special point in the retina, termed fovea, that is the point of highest visual acuity and is always directed at the object that one is paying attention to at the moment.

The eyeballs are held in their respective optical cavity by various ligaments, muscles and fascial expansions. Figure 2.4 shows the rectus tendon which is connected to one of the two pairs of muscles running to the skull called the rectus muscles. An additional pair of muscles, called oblique muscles, are responsible for rotating the eyeball in the orbit.

2.3.2 *The Retina*

The retina can be seen as an extension of the brain; it is responsible for transducing light into electrical nervous pulses, and for the early processing stages of neural visual signals. To understand retina function it is essential to first understand its anatomy [Kolb (2003)].

The retina is a circular disc with a diameter of approximately 4.2 mm and a thickness of 0.5 mm. It is composed of several layers of neurons that can be easily distinguished in a micrograph, like the one reproduced in Fig. 2.5. The darker layers, called the *nuclear* layers, contain the neuron cell bodies, while the pale layers, called the *plexiform* layers, contain their dendrites and axons. Figure 2.6 portrays a simplified schematic of the organization human retina, with the neurons composing each layer labeled.

The human retina has two types of photoreceptors: rods and cones, which are named after their shapes. The rods are long and thin, and are about 120 million in number (about 94.5% of all photoreceptors). The rods are very sensitive to light (capable of perceiving even a single photon), and enable *scotopic vision* – the visual response at lower orders of illuminance magnitude.

The cones are less numerous than the rods (totaling about 7 million), have a shorter and thicker shape, and are less sensitive to light. The cones provide the eye's *photopic vision* – the visual response at 5 to 6 orders of illuminance magnitude – and are responsible for color perception. There are three different types of cones in the human retina: the blue, green and red cones, corresponding to the visible light wavelength to which they are most sensitive. Figure 2.7 shows the spectral responses of the different types of photoreceptors present in the human retina. At intermediate levels of illuminance, both rods and cones are active, enabling *mesopic vision*.

The photoreceptors are not distributed uniformly in the retina. While the cones are almost exclusively concentrated in the fovea, where there are no rods, the rest of the retina is populated predominantly by rods. Fig-

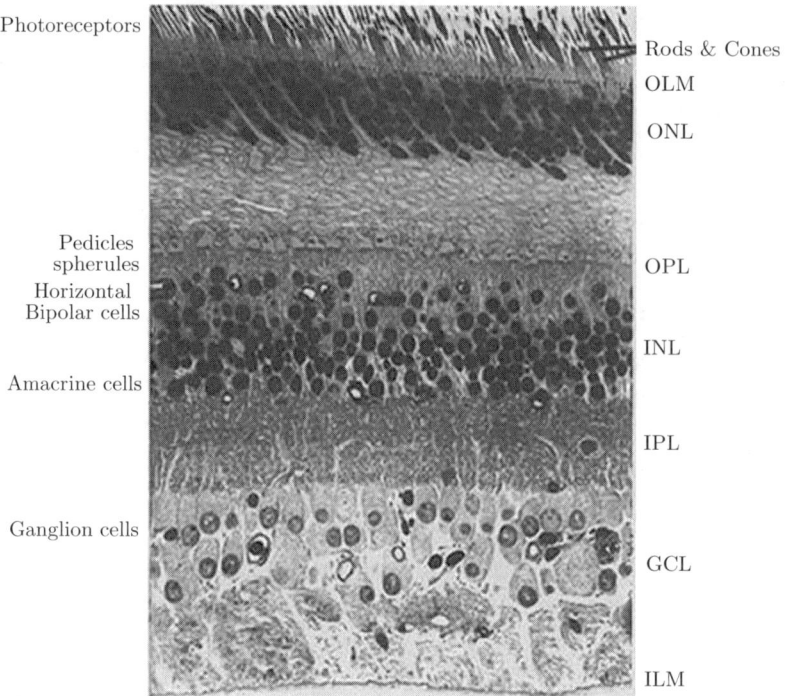

Fig. 2.5 Light micrograph of a vertical section through the retina (from [Kolb et al. (2002)]).

ure 2.8 displays a graphic with the photoreceptor distribution throughout the retina. Foveal cones are densely packed hexagonally, as is shown in the photograph of the cross section of the human fovea in Fig. 2.9(a). As distance from the fovea increases, the cones become larger and are packed less densely, like the photograph of the foveal periphery in Fig. 2.9(b) shows; the spaces between cones are filled by rod photoreceptors.

The fovea appears as a small dimple aligned with the visual axis (see Fig. 2.4), defining the operational center of the retina in bright light. Since in dim light an object focused on the fovea is not visible, at night we have to look to objects slightly sideways and is difficult to perceive color. The circular area around the fovea, with a diameter of approximately 6 mm, is the central retina; this region extends to the peripheral retina and further extends to the *ora serrata*, 21 mm from the center of the optic disc (see Fig. 2.4).

The light has to traverse all retinal layers to be sensed by the photore-

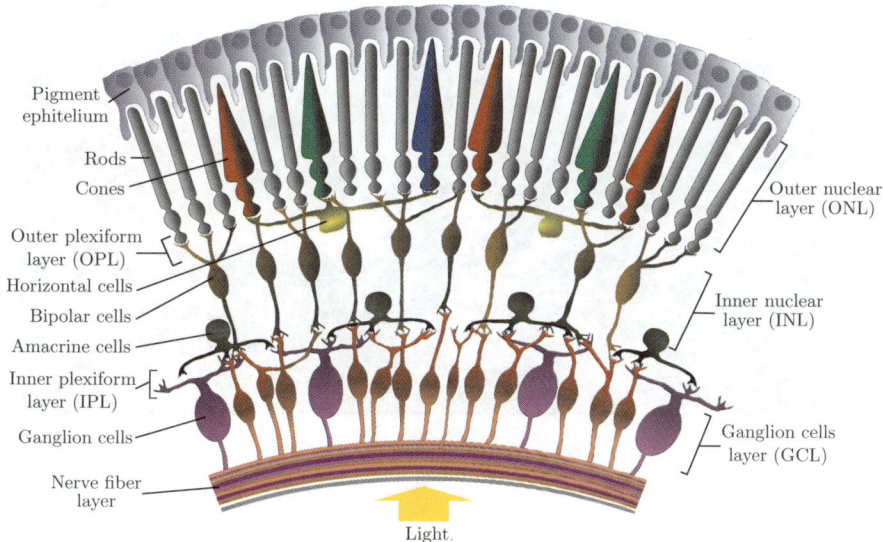

Fig. 2.6 Simplified schematic organization of the retina.

ceptors, as shown in Fig. 2.6, because photoreceptors must be in contact with the pigment epithelium (see Fig. 2.4). The exception to this rule occurs in the fovea, where nerves are pushed away so that the cones are directly exposed to light. Nevertheless, the photoreceptors are always in direct contact with the pigment epithelium because the pigment-bearing membranes of the photoreceptors have to be in contact with the epithelial layer, which provides a steady stream of several retinal molecules used to transduce light, like vitamin A and its aldehydes. These molecules sense light by changing their conformation in response to photons, and are then recycled back in the pigment epithelium [Rodieck (1998)].

The bottom layer of the retina in Fig. 2.5 is where the retina ganglion cells lie, and this is why it is called the ganglion cell layer (GCL). The axons coming from the retinal ganglion cells (RGCs) converge to one point where they leave the eye together and run to the brain, forming the optic nerve that also contains the blood vessels necessary for vascularization of the retina. This point is also called the blind spot, or optic disk, since there are no light receptive cells in this region (see Fig. 2.4). The existence of the blind spot is not perceived, because the blind spot of one eye is compensated by the other eye, and vice-versa.

The intermediate cell layers, namely the bipolar, horizontal and

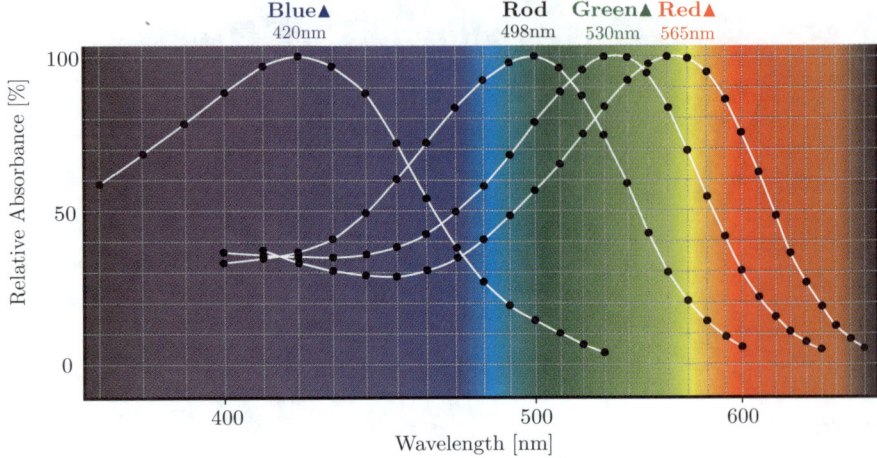

Fig. 2.7 Photoreceptors' spectral sensitivity.

amacrine cells, are responsible for processing the electrical stimuli coming from the different photoreceptors and enhancing the features relevant for the brain to extract information. These cells are organized in layers, as shown in the schematic of Fig. 2.6 and in Fig. 2.5, where the layers are labeled with the corresponding cell types. The photoreceptor cells' bodies are located in the outer nuclear layer (ONL). The inner nuclear layer (INL) contains the cell bodies of the bipolar, horizontal and amacrine cells. The

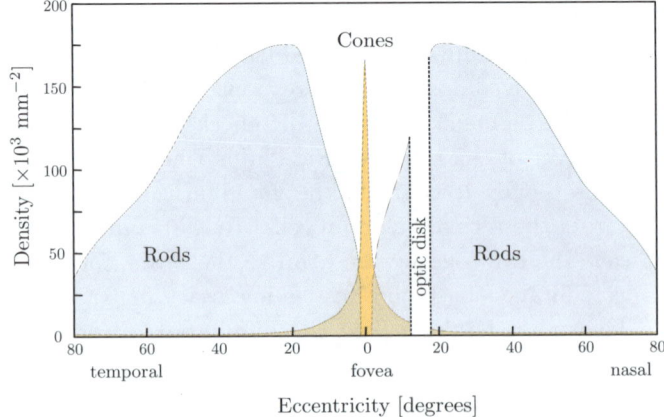

Fig. 2.8 Spatial distribution of photoreceptors along the retina.

outer plexiform layer (OPL) is located between these two neuron layers and contains the synaptic connections between the photoreceptors, bipolar, and horizontal cells. The GCL contains the ganglion cells, whose axons link to the brain. Between this layer and the inner nuclear layer is the inner plexiform layer (IPL), containing several kinds of synaptic contacts between the ganglion cells and the horizontally and vertically directed amacrine cells. Despite its complexity, we can obtain a general overview of the structure of the human retina by considering a simplified organization such as the one depicted in Fig. 2.6, which is a very basic view of the structure of the retina. In reality, the connections between the different types of cells are very intricate, as can be previewed from Fig. 2.5.

The central retina is thicker than the peripheral retina for two main reasons: the high density of photoreceptors present, mainly cones, and a greater density of cones connecting second order neurons located in the inner nuclear layer, called *cone bipolar cells*. There is also a greater number of ganglion cells in the ganglion cell layer of the central retina, which implies a greater number of synaptic interactions.

At the middle point of the central retina is the fovea, whose constitution is significantly different from the central and peripheral retina. In the foveal pit, the neuron layers are radially displaced and cones are compactly aligned in a hexagonal structure and exposed directly to light (see Fig. 2.9(a)). Around the foveal pit is the foveal rim, or *parofovea*. It is the thickest zone of the retina due to its six layers of ganglion cells that connect the central cones to the optic nerve. The foveal area, including the rim around the

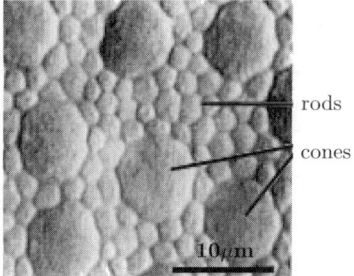

(a) Fovea cross section.　　(b) Peripheral fovea cross section.

Fig. 2.9 Photograph of the cross section of the fovea and of the foveal periphery: (a) cones densely packed in fovea, and (b) fovea periphery with bigger cone photoreceptors interleaved with rods (from [Curcio et al. (1990)]).

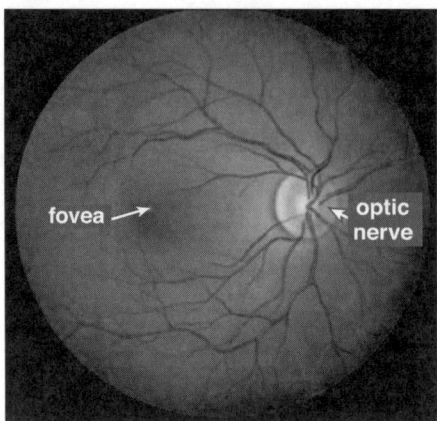

Fig. 2.10 Photograph of a human retina from an ophthalmoscope (from [Kolb et al. (2002)]).

parofovea, termed the *perifovea*, constitutes the *macula lutea*, which can be distinguished from the rest of the retina due to its yellow pigmentation (see Fig. 2.4). Figure 2.10 shows a photograph of the human retina taken with an ophthalmoscope, in which the fovea and the optic nerve are pointed out, and in which the blood vessels that vascularize the eye are visible.

Axons from both types of photoreceptors end in the OPL, where they establish visual pathways with neurons of the subsequent layers. To establish synaptic contacts with second order neurons, the cones end in a terminal called the *pedicle*, and the rods end in a terminal called the *spherule*, where the connections are established.

The cone pedicle ends in approximately 30 extensions associated with 30 triads of neurons. Each triad is composed of a central element, which is a dendritic terminal of a bipolar cell, and two lateral elements which are dendritic terminals of horizontal cells. There are also what are called *basal junctions*, consisting of different varieties of bipolar cell dendrites that form synaptic contacts under the surface of the cone pedicle. The rod spherule has two extensions; each one is associated with four, second order neurons. This group of four neurons consists of two horizontal cell axon terminals and two central rod bipolar cell dendrites.

In addition to the pathways between photoreceptors and second order neurons, there are also contacts between photoreceptors. These contacts are established between cones and cones and between cones and rods. The cone pedicles have small extensions from their bases that establish contacts with neighboring pedicles and spherules. These contacts are called *telodendria*. A rod can have 3 to 5 telodendrial contacts from neighboring cones, and cones can have as many as 10 telodendrial contacts to their neighboring rods. The exceptions are the cones that are sensitive to blue light, also called *S-cones*, which do not have many telodendria and have a second order specific kind of bipolar cell. These direct interactions between photoreceptors appear to degrade spatial resolution and color perception by mixing signals coming from different photoreceptors at different locations. One justification for telodendrial contacts is to allow rods to use neural pathways, devoted to both rods and cones, that are more rapid than the specific rod pathways.

2.3.3 How the Retina Operates

As mentioned earlier, the light must traverse all retinal layers before reaching the photoreceptors that lay at the back of the retina. The photoreceptors absorb photons that, by means of a biochemical reaction, are converted to an electrical signal capable of stimulating the forward neuronal layers of the retina. When the electrical stimuli arrive at the ganglion cell layer they are sent to the brain through the optic nerve as a sequence of stereotyped voltage pulses, called *action potentials* or *spikes*. (Figure 3.2 displays a typical spike waveform from a rabbit RGC).

The optic nerve contains about 10^6 optic nerve fibers, corresponding to extensions of the ganglion cell axons. Three different types of horizontal cells, 11 types of bipolar cells, at least 25 different types of amacrine cells have been morphologically distinguished and 18 types of morphologically different ganglion cells in the human retina have been identified [Kolb et al. (2002)].

When excited by light, the photoreceptors send a neurotransmitter through the vertical pathways of the retina. The horizontal and amacrine cells introduce excitatory or inhibitory signals in the retina's neural network, depending on their nature. A central concept in neuroscience is the notion of receptive field (RF). In the case of the retina, the RF can be characterized by the area on which light influences the neural response. Rods and cones react to light directly falling over them, so they have narrow receptive fields. Figure 2.11 represents a photoreceptor's receptive field. The rods can detect dim light and respond to relatively slow changes in luminance, while cones deal with bright light signals and can detect rapid light fluctuations. The process of image decomposition begins at the photoreceptor layer and continues in the first synapses of the visual pathway, existent between the photoreceptors and the bipolar cells.

Glutamate is a neurotransmitter that enables the transmission of the electrical neural signals from the photoreceptors through the bipolar cells to the ganglion cells. This neurotransmitter allows the establishment of electrical conductive channels between the axon of the presynaptic neuron and the dendrite of the postsynaptic neuron. Depending on the type of ion flow, the neurotransmitter effect can be excitatory or inhibitory through depolarization or hyperpolarization of the postsynaptic neuron, respectively. The horizontal and amacrine cells can send excitatory or inhibitory signals using several types of substances [Kolb et al. (2002)].

While in the dark, the rods and cones constantly release neurotransmitter; they cease releasing when excited by light. For example, when a green cone is in the dark, its internal potential is at rest, but when is illuminated by green light it becomes hyperpolarized – the electrical potential of its membrane gets more negative, stopping the neurotransmitter release over the duration of the light flash.

The distinct types of bipolar cells react differently to the neurotransmitter. Some bipolar cells can re-sensitize their glutamate receptors quickly and react to rapid changes in the visual signal by firing at a relatively high frequency, while others take more time to re-sensitize their glutamate

Fig. 2.11 Graphical representation of the narrow receptive field of a cone.

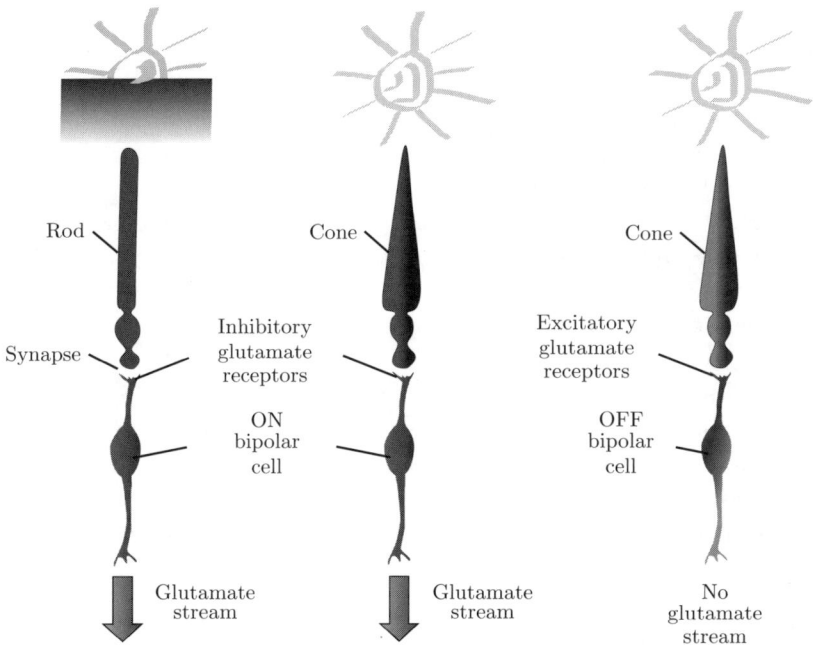

Fig. 2.12 Photoreceptor to bipolar cell connections. The ON-type bipolar cells (a,b); OFF-type bipolar cell (c).

receptors, so that they respond relatively slowly to the same amount of stimulus. On the other hand, some bipolar cells' receptors, which possess a hyperpolarizing receptive field, respond to glutamate by activating an OFF pathway, so that they detect dark images against a lighter background. Other bipolar cells possess inhibitory glutamate receptors that prevents the cell from firing when it receives the neurotransmitter; thus, the glutamate activates an ON pathway, and the cell detects a light image against a darker background and possesses a depolarizing receptive field. Figure 2.12 illustrates the different types of connections between photoreceptors and bipolar cells and the streams of information controlled by the glutamate molecule. A single bipolar cell receives input from a small number of cones and has a medium sized receptive field.

The parallel sets of visual channels of ON-type (detecting light areas on dark backgrounds) and of OFF-type (detecting dark areas on light backgrounds) are fundamental to sight, as vision depends on perceiving the contrast between an object and its background. Connections between ON-type bipolar cells and ON-type ganglion cells and between OFF-type bipolar cells and OFF-type ganglion cells occur in specific regions of the inner plexiform layer.

If the images were transmitted to the brain via the bipolar and ganglion cells alone, they would be grainy and blurry. The role of horizontal cells is to define the edges and enable the perception of fine details in an image. Each horizontal cell receives its input from several cones, and so its receptive field is large; the receptive field becomes even broader because the plasma membranes fuse with those of neighboring horizontal cells at gap junctions (see Fig. 2.13).

A single bipolar cell, with its ON or OFF light response, would carry a fairly blurry response to its ganglion cell. Horizontal cells add an opponent signal that is spatially constructive, giving the bipolar cell what is known as a *center surround organization* (see Fig. 2.13). The bipolar center signals can be either ON or OFF, and the horizontal cells add an OFF or ON surround signal through two different mechanisms - either directly or by sending feedback information to the cone photoreceptors, which then feed forward again to the adjacent connected bipolar cells. Figure 2.13 illustrates the information streams performed by the horizontal cells in conjunction with the bipolar cells.

Horizontal cells also receive feedback signals, from the inner plexiform layer, that influence its activity. The result is that horizontal cells modulate the photoreceptors' signals under different lighting conditions, allowing signaling to become less sensitive in bright light and more sensitive in dim light, as well as shaping the receptive field of the bipolar cells. Horizontal cells also make the bipolar cells' response color-coded through feedback circuits to cones.

Ganglion cells have a receptive field that is also organized in concentric circles. The amacrine-cell circuitry in the inner plexiform layer conveys additional information to the ganglion cells, thus sharpening the boundary between the center and surround areas in their receptive field even further than the horizontal cell input alone. There are two main types of ganglion cells in the human retina with ON centers and OFF centers that form the major output from the retina to the subsequent visual centers in the brain. The ON-center ganglion cells become active when a spot of light falls in the center of their receptive field and are inactivated when light falls on the field periphery. The OFF-center ganglion cells act in the opposite way; their activity increases when the periphery of their receptive field is lit and decreases when light falls on the center of the field. The horizontal cells convey antagonistic surround signals to bipolar cells, and consequently to

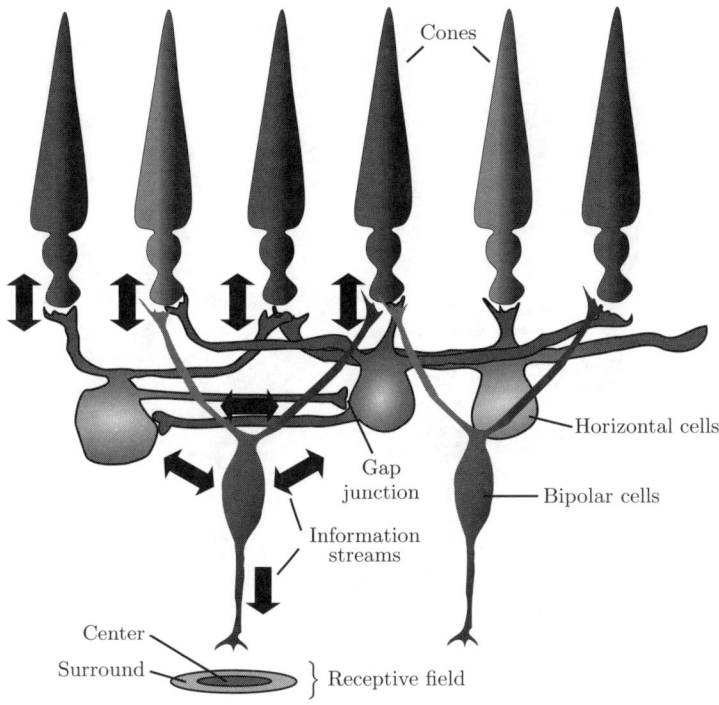

Fig. 2.13 Connections between photoreceptors, horizontal cells and bipolar cells, with the information streams indicated.

the ganglion cells. This kind of processing sharpens the boundaries of the images. The receptive fields of the ganglion cells can be modeled as the difference between two Gaussian functions, giving them a so called Mexican-hat shape. Figure 2.14 presents a sketch of the connections of an ON-type ganglion cell with its receptive field modeled as a difference between two Gaussian functions, represented at the bottom of the figure. An OFF-type ganglion cell is connected to OFF-type bipolar cells, and its receptive field would have a symmetric shape relative to the one in Fig. 2.14.

In contrast to the rest of the retina, the organization of the retinal cells in the fovea region contains midget ganglion cells, which possess tiny dendritic trees that are connected in a one-to-one ratio with midget bipolar cells. The channel from midget bipolar to midget ganglion cells carries the information from a single cone, thus relaying a point-to-point image from the fovea to the brain. Each red or green cone in the central fovea connects to two midget ganglion cells, so at any time each cone can either transmit a

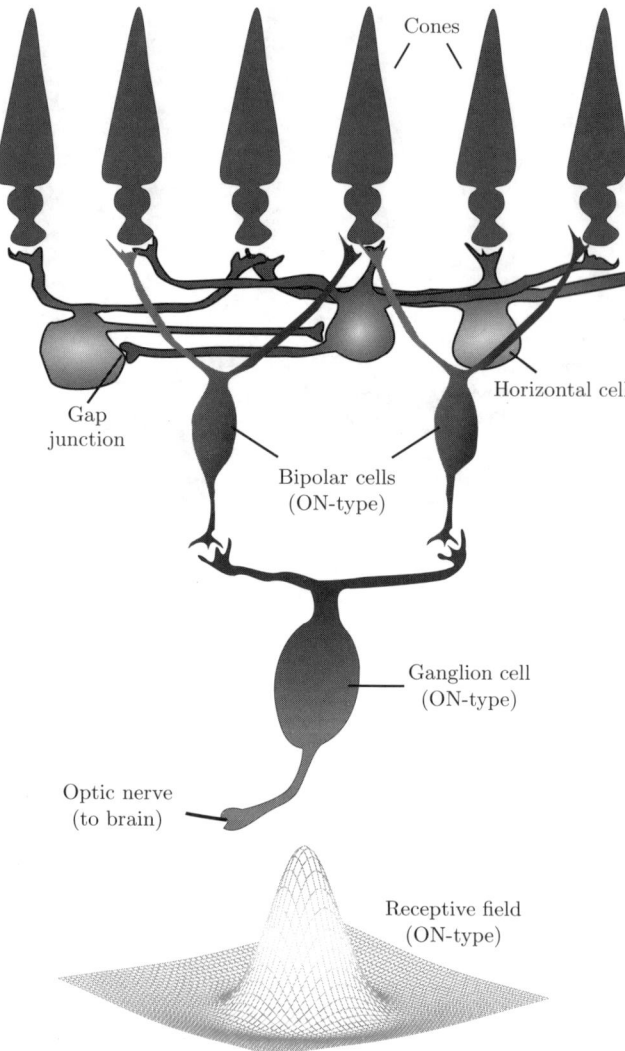

Fig. 2.14 Connections between bipolar cells and ganglion cells.

dark-on-light (OFF) signal or a light-on-dark (ON) message. The message that is sent to the brain carries both spatial and spectral information at the highest spatial resolution. As a result of being connected to only one cone, the receptive fields of the midget ganglion cells are much narrower than their counterparts on the rest of the retina. Blue cones are organized in a different way: they transmit information through a specific blue bipolar cell to a different type of ganglion cell, which can carry both a blue ON and a yellow OFF response.

At the bottom of the inner plexiform layer, just above the ganglion cell layer, are the amacrine cells. Among the several types of amacrine cells, there is a special type denominated AII. These cells receive information from the bipolar cells and transmit information to ganglion cells and bipolar and amacrine cells. They also provide interconnections between ON and OFF systems of bipolar and ganglion cells. Another important type of amacrine cell is designated A17. These amacrine cells are crucial in the pathways starting in rods. Whereas the cones connect directly to bipolar cells, and these to ganglion cells, the bipolar cells that receive their input from rods do not synapse directly to the ganglion cells. All bipolar cells connecting to rods are of the ON-type (see Figure 2.12), and use the AII and A17 amacrine cells to send signals to the ganglion cells. A single AII amacrine cell can be connected to as many as 30 rod bipolar cells and can transmit a depolarization signal both to ON cone bipolar cells and their ON ganglion cells, and to OFF cone bipolar cells and their OFF ganglion cells. Therefore, AII amacrine cells make it possible for rods to use the faster cone pathways. Figure 2.15 illustrates the connections made by an AII amacrine cell. The A17 amacrine cells also collect the signals from thousands of rod bipolar cells that are modulated, amplified, and transmitted to AII amacrine cells. Signal modulation and amplification allow the perception of very weak light signals, and consequently enable night vision.

There are several other types of amacrine cells that spread horizontally, interacting with hundreds of bipolar cells and many ganglion cells. They can even connect to neighboring amacrine cells through gap junctions, increasing their action radius and the speed with which signals can be transmitted across large areas of the retina. Another role played by the amacrine cells is in modulating the retinal response for different illuminance conditions. By liberating several kinds of neurotransmitters, they inhibit or reinforce the synaptic connections between the neuronal layers in the retina.

All these singularities of light processing in the retina suggest that a significant part of the construction of the visual images occurs in the retina. A quantitative analysis of the responses of each retina cell type permits the evaluation of each cell type's contribution to the ganglion cell response. The analysis performed in [Meister and Berry II (1999)] revealed that the photoreceptors, horizontal cells, and bipolar cells produce responses to light that are basically linear. On the contrary, under the same light stimulus

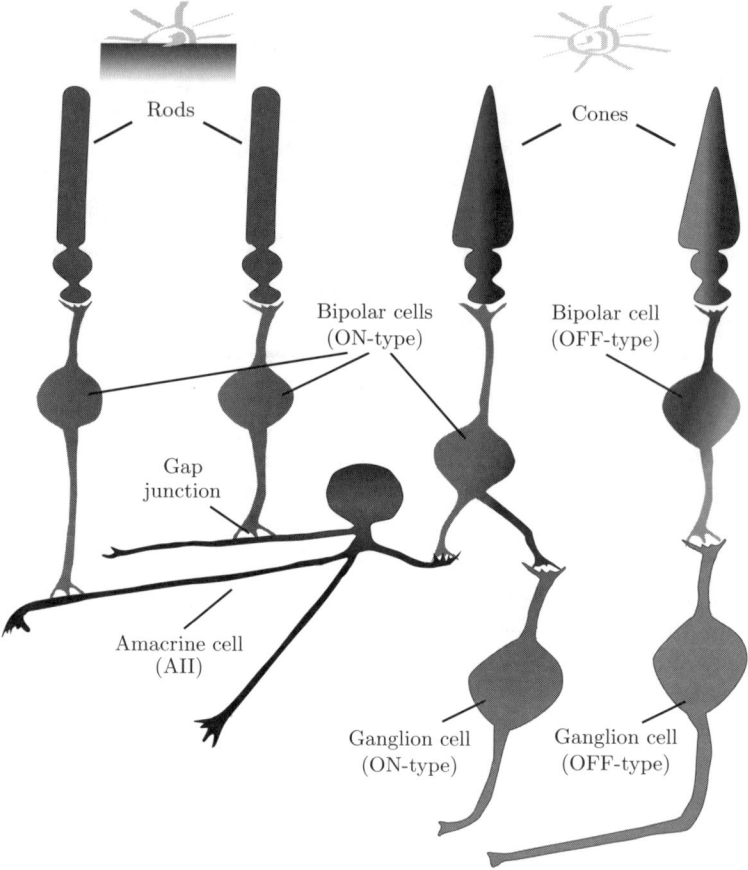

Fig. 2.15 AII amacrine cell connections.

conditions, the amacrine cells showed strong nonlinear distortions, which made it possible to distinguish between *sustained* and *transient* amacrine cells [Victor (1999)]. This kind of experiment allows the functional differences between classes of cells in the retina to be distinguished and the development of models in a cascading manner, giving biological meaning to each block in the signal processing pathway.

2.3.4 The Visual Pathway

In addition to the retina, the spatiotemporal processing of the illuminance pattern gathered by the eye continues to take place all along the visual pathway to the cerebral cortex. Figure 2.16 shows a schematic view of the visual pathways, from the retina to the cerebral cortex. A visual prosthesis can potentially be interfaced with the nervous system in different places along the visual pathway, with inherent advantages and disadvantages [Warren and Normann (2003)].

All the neurons along the visual pathway possess a receptive field. The receptive field construct can be expanded to include other characteristics of the visual stimuli, such as the shape, size, intensity, color, and location in the visual space, that drive the neuron to respond optimally. Another important issue is the mapping of the visual space to the neural space. This mapping is visuotopic, meaning that the neurons along the visual pathway are arranged such that their receptive fields form an organized and approximately linear map of the visual space. As a consequence, objects that are close together in the visual space evoke neural activity in nearby neurons in the brain. This implies that a rectangle in the visual space will result in neural activity in a similarly shaped arrangement of neurons in the visual centers, although this arrangement may be stretched along each axis, rotated, and/or warped. The external electrical stimulation of this ensemble of neurons would result in the perception of the outline of a rectangle. Figure 2.16 illustrates the visuotopic organization of the visual pathway, with points A and B in the retina mapping to points A and B in the visual cortex.

The visual pathway is a highly parallel signal processing system. This parallelization occurs along two principal parallel pathways: the M pathway (M for *magno*, or large) and the P pathway (P for *parvo*, or small). These distinct parallel pathways begin in the retina and extend through the visual pathway. The M and P pathways represent different features of an object placed on the visual space, such as where the object is located and what object it is, respectively [Warren and Normann (2003)].

2.3.4.1 *Optic Nerve and Tract*

The optic nerve contains the nerves running from the retina to the optic chiasm – the location where the nerves coming from the left and right eyes intersect (see Fig. 2.16). The nerves that go from the optic chiasm to the subcortical tracts compose the optic tract. The optic nerve has a length of approximately 50 mm and the optic tract is about 30 mm in length. It is uniquely composed of the axons of 1.2×10^6 retinal ganglion cells, and do not include neuron cell bodies. The optic nerve contains the axons from

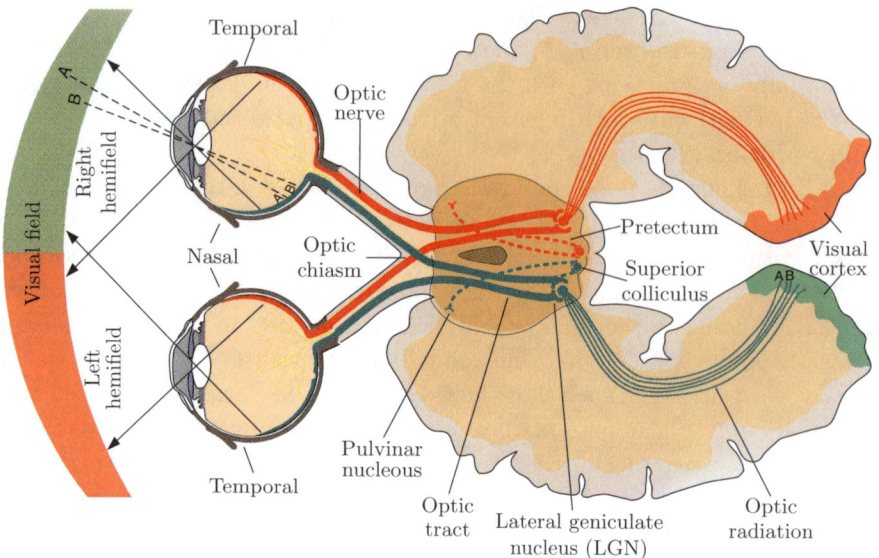

Fig. 2.16 Schematic cross section of the human brain with the visual pathways depicted.

the nasal visual field and from the temporal to the fovea visual field of a single eye. In the optic chiasm, the optic nerve axons are reorganized so that the optic tract almost exclusively contains the axons representing the contralateral visual hemifield. Besides the retinal ganglion cell axons, the optic nerve also contains an artery and vein, which irrigate the retina. Figure 2.16 shows a schematic cross section of the human brain illustrating the different visual fields and pathways.

The fibers of the optic nerve are also visuotopically organized, with the upper retina represented along its dorsal side, the central retina along the lateral side, and nasal visual field along the medial side. However, this visuotopic organization changes along the nerve [Warren and Normann (2003)]. Because the optic nerve and the optic tract are extensions of the retinal ganglion cell axons, their receptive fields have the same structure.

2.3.4.2 Subcortical structures

The axons coming from the retinal ganglion cells target three subcortical structures: the *superior colliculus*, the *pretectum*, and the lateral geniculate nucleus (LGN) (see Fig. 2.16). The superior colliculus and the pretectum are localized on the top of the midbrain, and are associated with the sac-

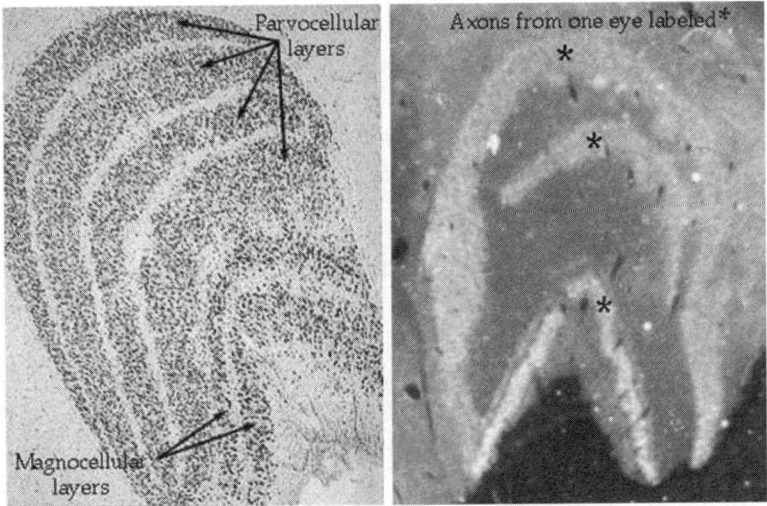

Fig. 2.17 LGN cell layers (from [Kolb et al. (2002)]).

cadic eye movements and pupillary reflexes, respectively. Neither of these structures is suited for a visual prosthesis implant or bioelectronic vision system.

The LGN is located on the ventral side of the *thalamus*, and its neurons are considered relay neurons: they receive the input from the retinal ganglion cells and pass it to the cortex. The volume of the LGN is small, $7 \times 7 \times 2$ mm, with six functionally independent cells *laminae*, stacked in a form of a distorted 'U' (see Fig. 2.17). Each lamina receives input from only one P or M pathway and from only one of the contralateral or ipsilateral eyes. Figure 2.17 displays a microscopic cross section of the LGN, in which the projections of the small P cells and large M cells from the two eyes to *parvocellular* and *magnocellular* layers of the LGN are indicated. Each eye projects its RGC axons to alternating layers, as indicated on the right side of the picture shown in Fig. 2.17.

The LGN laminae are visuotopically organized, and the visuotopic maps are registered between lamina. Half of the neurons of the LGN, representing half of its area, have receptive fields in the fovea and surrounding region [Warren and Normann (2003)]. Despite its difficult access, the neural organization of the LGN make it eligible for a neuroprosthesis implant and there is some experiments to demonstrate its feasibility [Pezaris and Reid (2007)].

2.3.4.3 Visual Cortex

The visual cortex is the final stage in the visual pathway. Even in the visual cortex, the M and P pathways continue to follow distinct processing paths. The majority of the LGN axons run to a cortex region, designated as the primary visual cortex because it is the first visual processing region at the cortical level, or cortex visual area 1 (V1), with dimensions within the range of 2500 − 3200 mm^2. This brain region is also known as the striate cortex. Visual processing takes place in subsequent areas of the cortex referred to as 2, 3, 4 and 5 (V2, V3, V4 and V5), reflecting their hierarchical relationships.

All these areas have a laminar structure containing six layers with a total thickness of approximately 2 mm from the *pia mater* (the more external region of the brain) to the white matter (the internal region of the brain). Processing of the visual information is performed in an upright manner, with information passing between neurons in a column from pia mater to white matter, and in an horizontal fashion, in which information is integrated across a number of columns. Layer 4 in area V1 is further subdivided into four layers: 4A, 4B, 4Cα and 4Cβ. Layers 4Cα and 4Cβ receive input from the M and P pathways, respectively, and layer 6 in V1 sends feedback to LGN [Grill-Spector and Malach (2004)].

Every area in the visual cortex is visuotopically organized. The visuotopic maps are registered between lamina so that all neurons in a column, from pia mater to white matter, have receptive fields in the same region of visual space. Half of the neurons in the visual area, representing half of the region size, have receptive fields in the fovea or surround region. The visual space corresponding to the foveal region is represented at the posterior part of V1.

The receptive fields in the visual cortex, with the exception of layer 4C of V1, which has receptive fields identical to the LGN, are more complex than those of previous centers. They still have subregions of preferred ON-type illuminance, lacking OFF-type illuminance, but their shape is no longer annular. They are somewhat elongated, giving preference to visual stimulus with a bar form oriented in a given direction. The visuotopic organization of the retinal input, preserved through all stages along the visual pathway, becomes less distinct in the cortical areas. A neuron can be driven by the contralateral eye, by the ipsilateral eye, or by both eyes. Nevertheless, nearby neurons tend to prefer the same orientation and receive information from the same eye. After the V1 area, the organization of the visual pathway becomes more complex and the optimal visual stimulus becomes less evident.

Only a small fraction of all the stages that exist along the visual pathway are suitable for the implantation of a neural interface to the nervous system. The main issues hindering implantation are related with retinotopic organization and with surgical access for interface implantation. The place for implanting the prosthesis is also related to the damage present in prior stages of the visual pathway. The area V1 is the most adequate for implanting a visual prosthesis at the visual cortex level [Warren and Normann (2003)].

2.4 Modeling the Retina

The development of a neural model for the retina has the potential to answer the following questions: "What code is used by the retina to codify visual information?" and "How is that code generated?" The answer to the former question would be like having a dictionary for this neural language, where we search the translation for a given visual stimuli. The answer to the latter one can be seen as the construction of the dictionary for this particular neural language - the retina neural code.

Next, some relevant peculiarities related with the retina neural code are explored, followed by a presentation of the taxonomy used for classifying the retina models and some frequently used processing blocks.

2.4.1 *The Retina Neural Code*

A central question in computational neuroscience is how the evoked potentials encode information, and the dual problem, given a spike train, how the stimulus that originated it can be reconstructed. These two processes are termed coding and decoding, respectively, and are intimately related [Rieke et al. (1997)]. In this context, the question is how the visual image is encoded into the spike trains that are sent to the brain, in a parallel fashion, through the optic nerve.

The brain does not directly interpret the light intensity pattern falling on the retina, but instead, it extracts information from the spike pattern carried by the optic nerve. The brain starts its processing with these spike sequences, and it sends information to motor neurons in the form of another sequence of spikes. The language of the brain is composed of sequences of spikes: the brain listens to spikes, uses spikes in its internal processing, and communicates with the external world using spikes [Rieke et al. (1997)].

The conversion of light into a spike pattern occurring in the retina has some distinctive characteristics that deserve to be pointed out. We will focus on the neural code employed by the ganglion cells of the retina for conveying visual information to the brain.

A remarkable characteristic of the retina is the amount of information compression that it performs, since the light is collected by 120 millions photoreceptors and the optic nerve has about 1.25 millions fibers (corresponding to the retinal ganglion cells axons), a relation of about 100 : 1. On average, we have 100 photoreceptors for 1 ganglion cell though this relation varies systematically along the retina. Another important feature is that the response of the retina depends on the temporal characteristics of the stimulus. For stationary stimuli the response vanishes, as well as above a certain frequency. Only within a given bandwidth does the retina fire spikes, behaving as a temporal bandpass filter. Moreover, the retina is not sensitive to the absolute illuminance, but instead is sensitive to its variation, the contrast.

Regarding the spatial composition of the stimuli, the retina output is directly related to the previously discussed notion of a receptive field – its response is maximal if the stimulus resembles the spatial form of the RF. This raised the existence of different types of retinal cells with ON or OFF centers, with OFF or ON surrounds, respectively, a retinal process known as lateral inhibition [Meister and Berry II (1999)].

Another important issue related to the retinal response is linearity. The retina shows a linear correspondence between the input stimulus and the firing rate for only some ganglion cells and under restricted conditions – the range of variations of light intensity must be small relative to its mean and cannot change too much over time. To model the nonlinearities of the retina response, a processing block denominated by contrast gain control (CGC) is frequently used [Meister and Berry II (1999)].

The retinal response changes depending on the illumination conditions. A consequence of this phenomenon, known as light adaptation, is that the capacity for distinguishing two different levels of illuminance depends of the mean illuminance level. The perceived contrast ΔC depends on the absolute luminance I, following the Weber-Fechner law [Wandell (1995)]:

$$\Delta C \propto \frac{\Delta I}{I}, \tag{2.1}$$

where ΔI is the variation in the luminance intensity. This is an important characteristic, since the illumination of the natural world changes several orders of magnitude during the day. Some ganglion cells transmit information about the absolute light illumination level to control the pupil and eye aperture, and the visual response is somewhat immune to the mean intensity level, coding only the contrast changes.

The spatial resolution is also affected by the light intensity conditions. In dim light conditions, the time response slows down and the ganglion cell integrates the incoming illuminance over a longer period of time. The receptive field gradually loses its antagonistic characteristics, reducing the surrounding area. This adaptation permits vision in dim light conditions, but at the cost of reduced temporal and spatial resolution. This adjustment of the stimulus-response relationship, dependent of the mean light level, takes place at different locations along the retina using distinct mechanisms [Meister and Berry II (1999)].

The retina not only adapts to the mean light intensity level, but also to its variance, or contrast. This much more finely tuned retinal response, denominated by contrast adaptation, represents a slow adjustment of the retinal code to the changes of the statistics of the visual image. The time course of contrast adaptation is different for a contrast increase compared to a contrast decrease, and it is supposed that contrast adaptation is not performed by individual photoreceptors, but that it is instead a collective process.

An old discussion, related not only to the retina neural code, but to the nervous system in general, relates to whether the neural code is a rate code or a time code [Nirenberg and Latham (2003); Eggermont (1998)]. The spike trains are so variable for different trials with the same stimulus, that some scientists claim that the brain retrieves the information about the stimulus from a spike train through the neuron firing rate $r(t)$. This firing rate is obtained by averaging the responses from the fires of many identical ganglion cells [Meister and Berry II (1999)], and composes a rate code. On the other extreme are the scientists who advocate that the precise time occurrence of a spike, like the time interval between spikes convey relevant information about a given stimulus; all the time characteristics associated with the neural function $\rho(t)$ are important [Victor (1999)] and compose the time-code [1]. This point of view is reinforced by the fact that analyzing a trial response to a given stimulus involves discrete events in time, and not a continuous firing, and that each event is well described by the time

---
[1]The firing rate $r(t)$ and neural function $\rho(t)$ are two distinct ways of characterizing a neuron response, and are defined in the next chapter.

of the first spike and the total number of spikes in the event [Berry II and Meister (1998); Uzzell and Chichilnisky (2004)].

Another question is whether the neural code of the retina can be viewed as an individual process, where each individual neuron cell acts independently, or as a population process. Putting it another way, whether the firing of a neuron is dependent on the response of its neighboring neurons, which would imply that correlations between firing patterns generated by different neurons convey information [Eggermont (1998)]. For the case of retinal ganglion cells, several studies suggest that the neural coding at the retinal level is essentially an individual process, in which the population coding is responsible for carrying only a minor part of the information transmitted, and when the correlation between spike trains is not taken into account, more than 90% of the information about the stimulus can be retrieved [Nirenberg et al. (2001)].

2.4.2 Classification of Retina Models

Several approaches have been proposed to modulate different features of the retinal processing. Some relevant examples are the general Weber-Fechner law, which interprets the light adaptation that occurs in the retina, or the edge detector proposed by Marr, known as a Laplacian of Gaussian (LoG) [Lim (1990)], inspired in the ability of the retina to devise contours in images. While these models provide some insight into the processes occurring in the visual system, they are not able to mimic the fine details of retinal processing and do not provide a sequence of evoked potentials, as desired.

As presented in the previous sections of this chapter, the retina is organized in several layers by discrete processing units, such as photoreceptors, bipolar and horizontal cells, amacrine cells and ganglion cells. Despite this fact, many retina models are developed using a continuous formalism, bypassing processes such as the discrete-space sampling done by the photoreceptor mosaic, but providing a mathematical description in a closed form.

Looking for a taxonomy to classify retina models, one suggestion mentioned previously (in Chap. 1) is to separate the models into two groups: the functional models and the structural models [Wulf (2001)]. Models can also be classified using other characteristics, such as rate-code versus time-code models, depending on whether the output is the firing rate or a discrete sequence of firing events, where the time occurrence of each spike is taken into account.

Within the two great classes of retina models, there are other characteristics that can be used to distinguish the models, such as: the type of the mathematical description used for the model (differential equations, discrete equations), relation grade with the functional anatomy of the retina, and grade of reproduction of the experimental data (e.g. experimental data that cannot be reproduced) [Wulf (2001)].

2.4.2.1 *Functional Models*

A functional model attempts to mimic the functions of the retina as a black box, mapping its input into its output.

These models describe the spatiotemporal receptive fields of the ganglion cell by a set of equations. Typically, they involve one input layer, where light enters, and one output layer, that furnishes the model output; often the input and output layers are seen as the photoreceptor and ganglion cell layers, respectively, with the other interneurons connections not taken explicitly into account [Wulf (2001)].

Functional models can be composed of different blocks, performing different types of processing. A functional model can have only a spatial block, where the temporal processing of the retina is set aside, or can be complemented by a temporal block that models the temporal processing of the retina. It can also be composed of only a temporal block, where the spatial processing is set aside. A functional model can also be composed of a single spatiotemporal block, where the separation between the spatial and temporal processing cannot be made.

2.4.2.2 *Structural Models*

A structural model attempts to model each individual retina cell layer, and the cascade of sub-models for each retinal layer provides a complete retina model. This class of models include models of only a given retinal cell type, and models of an interconnected subset of retinal cells, resembling the anatomical structure of the retina.

Frequently, the model for each cell is based on the Hodgkin-Huxley model of the neuron [Dayan and Abbot (2001)]. The neuron cells are modeled in terms of electric circuit components that simulate the ion channels and ion currents flowing through the cell membrane [Wulf (2001)].

2.5 Conclusions and Further Reading

The human visual system is very complex, involving different areas of the nervous system, and comprising several distinct processes. For a general overview of the human visual system refer to [Wandell (1995)]. A delightful and colorful book about the first steps in the visual process that relates the physical with the physiological processes is [Rodieck (1998)]. Another important reference on the subject is [Dowling (1987)].

Two references related to the subject treated in this chapter are [Kolb (2003); Kolb et al. (2002)]. In particular, [Kolb et al. (2002)] is a major source of information about the human vision system, with the additional value that it is periodically updated and further complemented with comprehensive material from different researchers. It contains up-to-date information about the human visual system, that we have only touched on in this introductory overview. The review in [Grill-Spector and Malach (2004)] contains relevant information about the organization of the human visual system in general, and about the visual cortex in particular, addressing the retinotopic organization of the visual system. For a general overview of the processes involved in the neural processing, with some examples concerning the visual system, see [Dayan and Abbot (2001)]. The understanding of retinal vision processes is far from complete. Every day brings to light additional findings about this amazing sense, where the division between the physiological and psychological phenomena is sometimes hard to establish [Werblin and Roska (2007)].

Exercises

2.1. *List the main components of the human eye with their principal functions.*

2.2. *Referring to the photoreceptors present in the human retina:*

2.2.1 *What are the main characteristics of both types of photoreceptors?*
2.2.2 *What is their main location based on their relative quantities?*
2.2.3 *Investigate the process of light transducing at the photoreceptor's layer of the retina.*

2.3. *Present the different types of bipolar cells, describing their importance* to vision.

2.4. Present the connections between photoreceptors and bipolar, horizontal, and ganglion cells, relating the way that images are transmitted to the brain through the optical nerve.

2.5. Compare the organization of the retinal cells in the fovea against its periphery. Where is the spatial resolution higher? Why?

2.6. Describe the mechanisms, in terms of the retina structure, used to decrease the retina response time. Where are these mechanisms more effective?

2.7. Besides the eye, what are the other main centers of the visual pathway? Describe their organization and enumerate their main characteristics.

2.8. Describe the main advantages and disadvantages of all the centers along the visual pathway with respect to interfacing a visual prosthesis.

2.9. What is the importance of visuotopic organization in the development of a visual prosthesis? Describe the visuotopic organization along the visual pathway, from retina to the visual cortex.

Chapter 3

Characterization of the Neural Response

3.1 Introduction

This chapter presents several quantities and mathematical tools that allow a description of the neural response. For an engineering audience, the literature on computational neuroscience in general, and for retina modeling in particular, is full of somewhat unfamiliar terms, definitions, notions and concepts. Despite the fact that, in the essence, many of the concepts are coincident (so that their main differences reside primarily in the terminology), sometimes there are some slight changes in the mathematical definitions. In other cases, however, completely new concepts exist.

In the early days of experimental neuroscience, the neural response was measured by counting the number of spikes occurring in a given time window applied at the onset of the stimulus presentation. In the experimental setups in use today, the retina is repeatedly stimulated with the same stimulus, and the acquired results are averaged over the experimental trials.

We start by describing the main issues in the stimulation and recording of the responses of the retina. A first look at the resulting spike sequences reveals that the neurons' responses are not equal for the same stimulus. Instead, the responses show some variability from trial to trial, meaning that the neural behavior has a certain level of randomness. This randomness implies that the neural code does not possess a one-to-one correspondence, but instead the same stimulus triggers different neural responses despite the fact that neurons have a certain degree of similarity. To uncover the neural code, it is essential to quantify the degree of randomness in the neural responses. Thus, we will present the main tools from probability and stochastic process theory used to describe neural responses.

The remainder of the chapter is devoted to presenting the methods

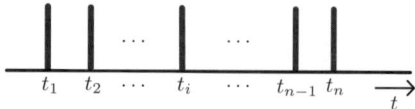

Fig. 3.1 Representation of a spike train.

and quantities used to analyze and characterize the neural response of a nervous retinal cell, so that we can quantify and thus reproduce its behavior. Although the methods and tools presented in the subsequent sections could be applied to the analysis and characterization of the responses of any neural cell, we will specifically target the neural response of the retinal cells.

3.2 Spikes: The Essence of the Neural Code

Regardless of the input stimulus type – continuous or discrete – the neural output is always a discrete sequence of voltage pulses, also called evoked potentials and briefly termed *spikes*, that are positioned at time instants $\{t_i\}$, with $i = 1, \cdots, n$, in the train, like the representation in Fig. 3.1. The waveform of the evoked potentials has a stereotyped shape for a given class of neural cells. Moreover, it is common practice to classify the ganglion cells by the shape of their evoked potentials [Wandell (1995)]. Figure 3.2 shows the spike waveform of a rabbit OFF-type RGC, sampled with a frequency $f_s = 30$ kHz, with the sampled points dotted.

Because of the stereotypical form of the spikes, despite the fact that their time lengths, amplitudes, and shapes have slight variations, the information carried to the brain must be encoded in the spikes' temporal occurrence instants. With regard to this fact, a spike train can be represented by a time series of equal amplitude bars, with a bar located at every instant where a spike occurs. Figure 3.3 shows a graphical representation of a spike train segment from the response of a rabbit transient brisk OFF-type ganglion cell when excited with a Gaussian random stimulus [Keat et al. (2001)].

A common, and not completely answered, question is: how does a spike train represent the sensory input, the internal states of the brain and the subsequent motor control? The answer to this question can be viewed as the search for a translation dictionary for this peculiar language [Meister and Berry II (1999)]. An interesting, and funny, illustration used to introduce

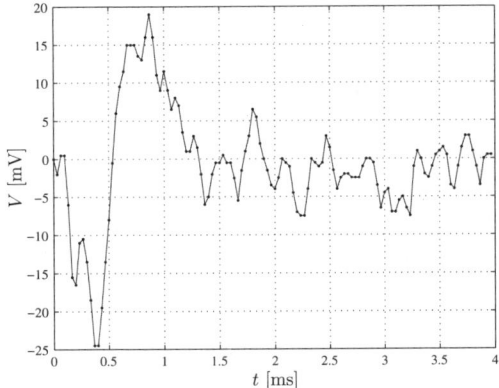

Fig. 3.2   Spike waveform of a ganglion cell from a rabbit's retina.

this problem is the concept of the *homunculus* [Rieke *et al.* (1997)]. The homunculus is a metaphor in which a little man, placed in the brain, receives the spike trains produced by the sensory organs in response to stimuli, and tries to figure out what stimulus the organism is sensing. After decoding the stimuli from the spike trains, the homunculus has to generate a spike sequence to communicate with the organism's members as a reaction to the environmental changes. This metaphor encompasses the two processes involved in the neural code: coding and decoding. The homunculus receives the sensory data encoded as a spike train, decodes it in order to perceive the stimulus, and has to encode the response again as a spike train to communicate its reaction.

In the development of a retina model, the main goal is to understand how stimuli are encoded. This process is intimately related with the decoding process, because when a stimulus is classified, by decoding the spike train it generates, the underlying coding mechanism is revealed.

A current topic in the research community is the identification of the

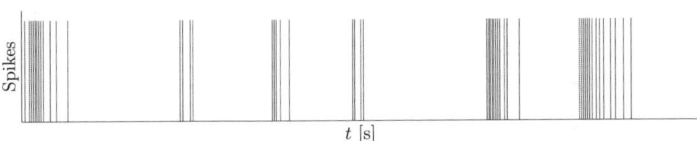

Fig. 3.3   Neuronal response function of a retinal ganglion cell.

relevant characteristics of the spike train that convey information [Eggermont (1998)]. Owing to the variability of the neurons' responses to the same stimulus, several researchers claim that the only significant feature in the spike train is the firing rate, and not the individual time instants of each spike occurrence. This viewpoint is usually called the *rate coding* approach [Berry II and Meister (1998)].

On the other hand, several physiologists claim, based on some recent studies, that by stimulating the retinal ganglion cells repeatedly, with the same visual stimulus, the cells' responses are quite regular with only a limited variability [Uzzell and Chichilnisky (2004)]. Thus, some recent retina models group the set of spikes from the neuron's responses into firing events, and assume that the time occurrence of the first spike, the time interval between the first and second spike, and the total number of spikes within each event are relevant to encode information [Berry et al. (1997a); de Ruyter van Steveninck and Bialek (1988)]. This perspective of the neural code is termed a *time coding* approach [Berry II and Meister (1998)]. Thus, in the time coding approach, the precise time relations between the spikes from the same neuron are considered to be meaningful.

Another current debate relates to the information conveyed by a retinal ganglion cell; whether it is independent of nearby cells, or if the information coded by the cell population response is relevant. Some studies have shown that 90% of the information is coded by the ganglion cell alone, and only 10% of the information is encoded in the population response [Nirenberg et al. (2001)].

3.2.1 *Retina Stimulation and Responses Recording*

The retina data have been obtained by sampling the responses of stimulated animals RGCs. The most commonly used retinas for data recording have provenance from different animals, mainly vertebrates such as rabbits, salamanders, turtles, monkeys, and even humans. Some invertebrates, like the blowfly, are also used [Keat et al. (2001); Berry II et al. (1999); Chichilnisky (2001); de Ruyter van Steveninck and Bialek (1988)].

The spike's voltage course can be recorded extracellularly or intracellularly. Figure 3.4 depicts the possible different locations of the recording electrode to measure neural activity. The signals recorded intracellularly are stronger and low-noise, while the extracellular recordings are weaker and subject to noise from neighboring cells [Dayan and Abbot (2001)]. Usually, an extracellular electrode collects electrical signals originating on

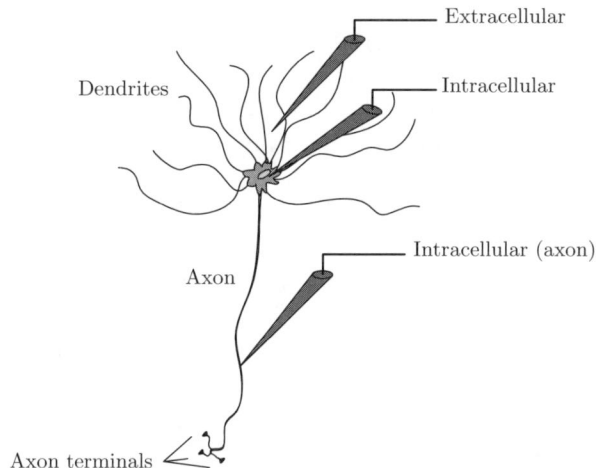

Fig. 3.4 Cellular recording of neuronal signals.

different neurons, which causes some difficulties in its analysis, particularly for classifying the cells according to the waveforms associated with spikes. Thus, if the recorded signals come from different neural cells of the same type, it is impossible to perform their classification, and so it is assumed that the spikes were produced by the same cell.

Currently, in the most common apparatus, the photoreceptor layer of the retina is excited with visual light patterns, and the neural responses coming from the RGCs are measured with an array of electrodes (see for example [Meister *et al.* (1994)]). Another, less common, alternative is to directly stimulate the retina with electrical currents induced through the electrodes, like in [Grumeta *et al.* (2000)].

The visual stimuli used to stimulate the retina RGC can have many variants and the choice of stimulus is usually tied to the particular retinal response that we want to characterize. In its most general form, a visual stimulus can have a spatial, temporal and light wavelength dependency: $s(x, y, t, \lambda)$, or $s(\mathbf{r}, t, \lambda)$ where $\mathbf{r}$ is the column vector $\mathbf{r} = [x\, y]^T$. Usually, the analysis of the retina's neuronal response is restricted to a particular subset of this general ensemble. The visual stimuli can be divided into two main categories in terms of their spatial and temporal variability. The retina's response is also commonly investigated for a particular wavelength, so that the dependence of the stimulus on the light wavelength can be dropped and we can focus only on the image luminance. The analysis of the chromaticity response of the retina is made by varying the stimulus wavelength.

In terms of their time variation, the stimuli are usually classified as deterministic or stochastic. In terms of the spatial behavior of stimuli, they can be organized as full-field or uniform type, meaning that they are spatially constant and do not convey any spatial information, or they can carry spatial information by changing spatially. The deterministic stimuli can have a closed mathematical description, such as the ON-OFF-type, or can be composed of simple patterns changing in time, such as a moving bar drifting along the visual field with a given speed. The stochastic stimuli can change both spatially and temporally in a random fashion. Figure 3.5 and Fig. 3.6 display the stimulus, and the retina responses, for the rabbit and salamander data, respectively, used along the book.

The simplest subset, called the binary subset, comprises spatially uniform stimuli with only two different intensity levels, like that illustrated in Fig. 3.7(a) (see also Fig. 3.5(a)). If the time duration of the presentation of each intensity level is constant, we have a periodic waveform and the stimulus is called ON-OFF. Otherwise, if the time period for each intensity level varies in a random way, it is called a flash stimulus.

Still, in the spatially uniform ensemble, a different stimulus intensity can be chosen for each plate, which is called a random stimulus and is shown in Fig. 3.7(b) (see also Fig. 3.6(a)). The light intensity levels can be tabulated previously or can be obtained by randomly sampling a probability distribution (like a Gaussian distribution, for example [Keat et al. (2001)]). This kind of stimulus is generally called white noise. A software library with a wide variety of types of stimuli and functionalities available to drive and control the images produced in a computer can be found at [Vision Egg (2007)].

In the ensemble of spatially non-uniform stimuli, there are many variants, ranging from simple bars to films of natural scenes. Figure 3.12(a) represents simple horizontal and vertical bars that sweep the visual field at a predetermined velocity. The bars can be viewed as a particular case of the more general stimulus called sinusoidal grating displayed in Fig. 3.8(a). The sinusoidal grating has the mathematical expression:

$$s_{\text{sinu}}(x, y, t) = A \cos\left(kx \cos\theta + ky \sin\theta\right) \cos(\omega t) , \qquad (3.1)$$

where the parameter k represents the spatial frequency, in radians per meter (rad·m^{-1}), so that $\lambda = 2\pi/k$ is the wavelength of the grating. The parameter θ controls the direction of stimulus movement. We can see Fig. 3.8(a) as if we had taken a photograph of the stimulus at a given

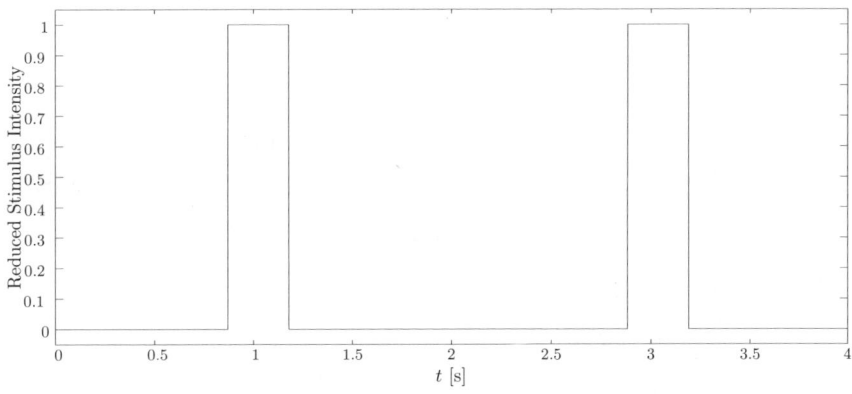

(a) Flash stimulus (or ON-OFF).

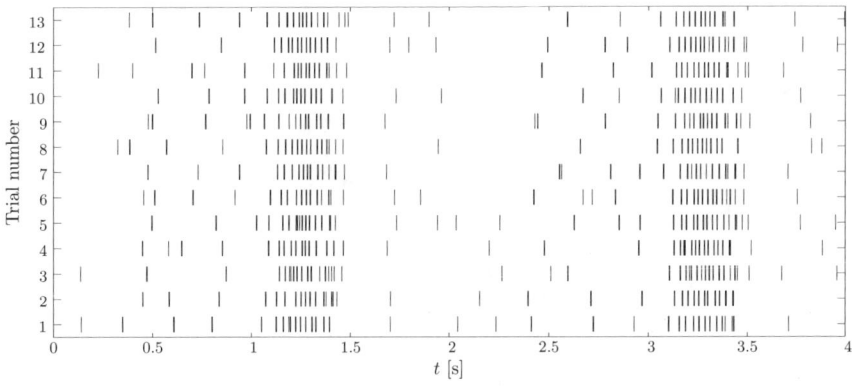

(b) Rabbit type-ON RGC spike trains.

Fig. 3.5 Rabbit type-ON RGC responses for a ON-OFF full-field stimulus ($T_s = 1$ ms).

time instant, but the stimulus amplitude in a given position (x, y) changes sinusoidally from the maximum $A \cos(kx \cos\theta + ky \sin\theta)$, to the minimum $-A \cos(kx \cos(\theta) + ky \sin(\theta))$ according to the temporal frequency ω, in radians per second (rad·s^{-1}), so that $T = 2\pi/\omega$ is the period of the sinusoidal modulation wave. Figure 3.8(b) shows the time evolution of the stimulus amplitude at a given point (x_0, y_0) where $s_0 = s(x_0, y_0, 0)$. To model moving bars from the sinusoidal grating, we simply have to consider that whenever $s_{\text{sinu}}(x, y, t)$ is positive it is equal to A, and when it is negative it is equal to $-A$. By introducing the signum function, sgn(x), defined

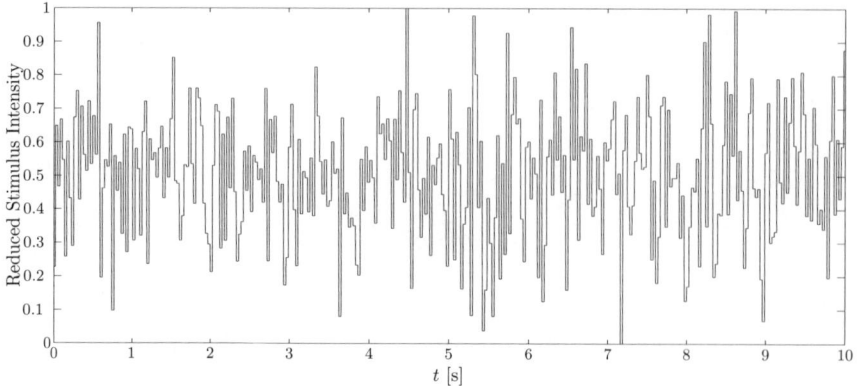

(a) White noise stimulus.

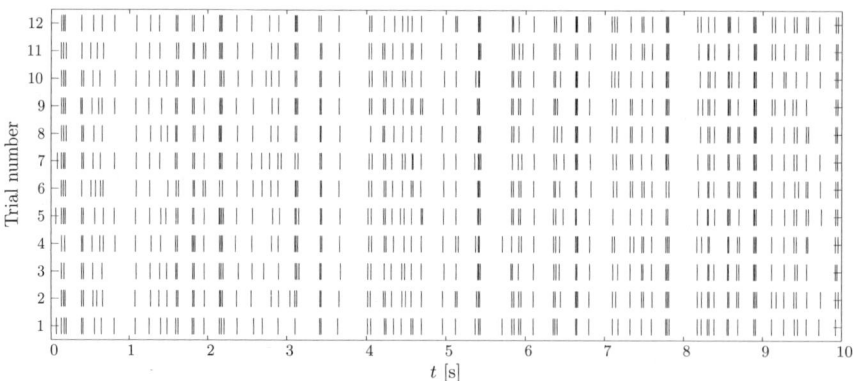

(b) Salamander type-ON RGC spike trains.

Fig. 3.6 Salamander type-ON RGC responses for sampled white-noise full-field stimulus ($T_s = 1$ ms).

by:

$$\text{sgn}(x) = \begin{cases} -1 & x < 0 \\ 0 & x = 0 \\ 1 & x > 0 \end{cases}. \qquad (3.2)$$

The moving bars can be expressed based on Eq. (3.1) like:

$$s_{\text{bars}}(x, y, t) = A \, \text{sgn}\left(\cos\left(kx \cos\theta + ky \sin\theta\right)\right) \cos(\omega t). \qquad (3.3)$$

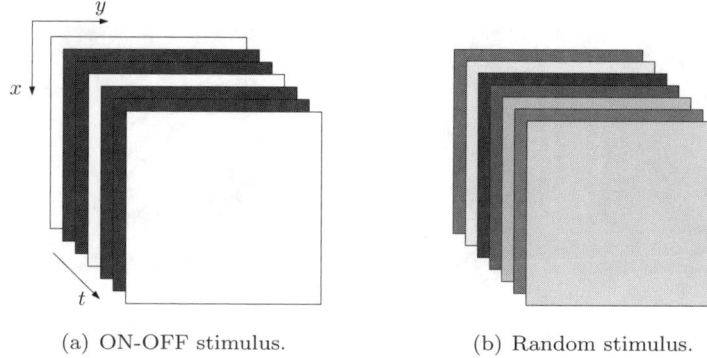

(a) ON-OFF stimulus. (b) Random stimulus.

Fig. 3.7 Spatially uniform visual stimuli.

Another set of commonly used visual stimuli are based on the Gabor functions. These kinds of stimuli are predominantly used in the characterization of the RF of ganglion cells, and in some simple RGCs where the structure of their RF has a Gabor function shape [Dayan and Abbot

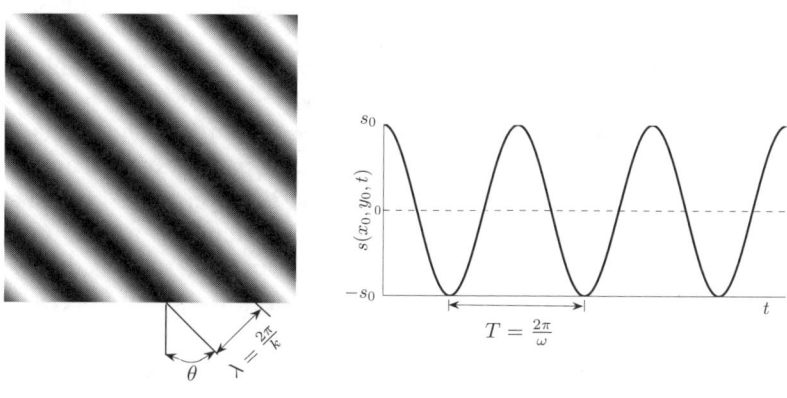

(a) Sinusoidal grating. (b) Sinusoidal modulation.

Fig. 3.8 Spatially non-uniform visual stimuli.

(a) Horizontal Gabor function.   (b) Rotated Gabor function.

Fig. 3.9   Spatially nonuniform Gabor functions.

(2001)]. The expression for the Gabor function is:

$$s_{\text{Gabor}}(x, y, t) = \frac{A}{2\pi\sigma_x\sigma_x} e^{-\left(\frac{x^2}{2\sigma_x^2} + \frac{y^2}{2\sigma_y^2}\right)} \cos\left(kx\cos\theta + ky\sin\theta\right)\cos(\omega t) , \quad (3.4)$$

and it can be seen as a two-dimensional, independent Gaussian function, with the span controlled by the parameter σ_x in the x-direction, and by σ_y in the y-direction, multiplied by a sinusoidal grating. Thus, the two-dimensional Gaussian function is modulated in space by a sinusoidal function with vector number k (the spatial wavelength), and is also modulated in time with a sinusoidal function with temporal frequency ω. Figure 3.9 shows an image of a horizontal and of a rotated Gabor function. The dark and lighter bands spatially move from image to image in time.

The stochastic stimulus class comprises stimuli whose spatial and temporal information are random. Figure 3.10 shows two such kinds of stimuli. Figure 3.10(a) shows a frame of white dots chosen randomly against a dark background, while in Fig. 3.10(b) each dot has a variable amplitude obtained by randomly sampling a Gaussian distribution. This type of stimulus is commonly referred to as a white noise stimulus, and the analysis of the respective neural responses is recognized to have several interesting features such as: exploration of a larger portion of the input space; insensitivity to the strong adaptation of the retina to a particular deterministic stimuli; and a receptive field estimation that is more robust to the fluctuations in the responsiveness of the neuron [Pillow and Simoncelli (2003)]. A whole set of analysis tools, particularly white noise analysis, are based in the study of the retinal responses to such stimuli [Westwick and Kearney (2003); Rieke et al. (1997); Rugh (1981)]

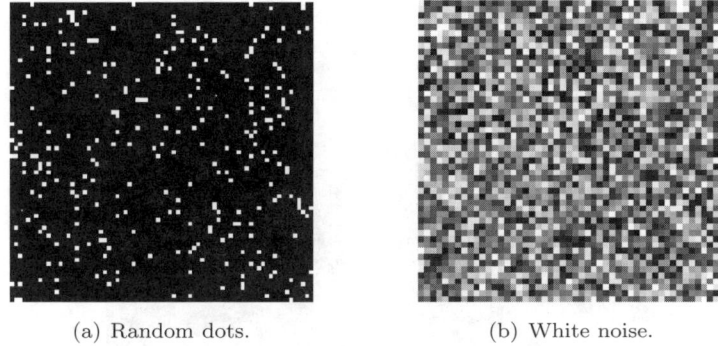

(a) Random dots. (b) White noise.

Fig. 3.10 Stimuli with spatial and temporal modulation.

In the current experimental apparatus, a computer monitor, with the cathode ray tube (CRT) or thin-film transistor (TFT) technology, is used to present the stimulus that is guided through an optical system to stimulate the retina. Figure 3.11 displays a more complex stimulus, where each frame is divided into small squares that, in the limit case, can be the image picture elements, whose RGB intensities are driven by randomly sampling a Gaussian distribution. In order to display the stimulus on a computer monitor, the stimulus must be discretized in time. The stimulus in Fig. 3.11 is discretized, where $t_{i+1} = t_i + \Delta t$ so that Δt is the time bin width, that establishes the sampling period of the stimuli, T_s. The images displayed by modern computer monitors are digital, so each frame is composed of a set of small picture elements (pixels), which correspond to an inherent fine-grained spatial sampling.

The retina has a response to the light stimulus intensity spanning several orders of magnitude, from a single photon to an influx of several millions of photons per second. The retina, similarly to other sensory organs, performs stimulus intensity compression, adapting to the mean level of the stimulus, and senses only deviations from the stimulus mean following the Weber-Fechner law (see Sec. 2.4.1). In order to simplify the study of the responses of retinal ganglion cells, the stimuli can be described only in terms of their var around their mean level. For a continuous-time stimulus this means that:

$$\frac{1}{T} \int_0^T s(t)dt = 0 , \qquad (3.5)$$

where T is the stimulus time duration. For a discrete time stimulus used

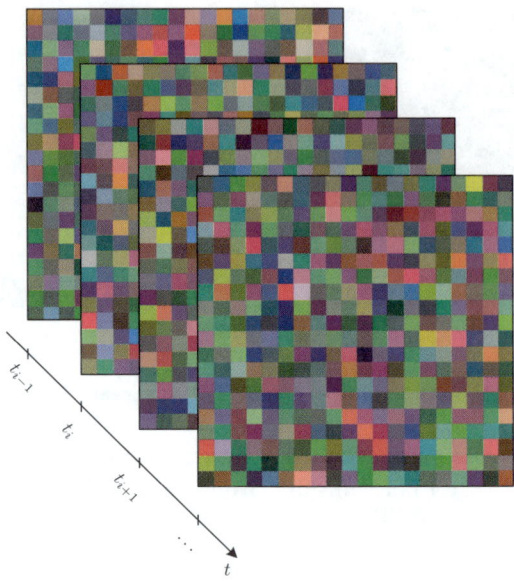

Fig. 3.11 Gaussian white noise stimulus sequence with spatial, temporal and chromatic variation.

in signal processing systems, the mean value of the stimulus is:

$$\frac{1}{N}\sum_{n=1}^{N} s[n] = 0 \;. \tag{3.6}$$

We can obtain the stimulus property expressed in Eq. (3.6) by subtracting its average value. Departing from a stimulus where Eq. (3.6) applies, a value can be added to the stimulus value in order to establish the desired mean level for stimulating the retina.

Next, we will provide an overview of the apparatus used at the University Miguel Hernandez to gather retina data from several types of vertebrates, including humans. This experimental apparatus is composed of three main parts: a stimulation system, a retina holder and positioning device, and a data acquisition and recording system. Figure 3.14 displays the experimental apparatus for retina data acquisition.

The most important component of the stimulation equipment is a computer with a graphics card capable of driving two monitors: one used for the experimental stimulus control and the other to present the stimuli. The displayed stimulus is reflected by a mirror and deflected by an optic prism in order to stimulate the retina positioned in an appropriate holder.

Characterization of the Neural Response

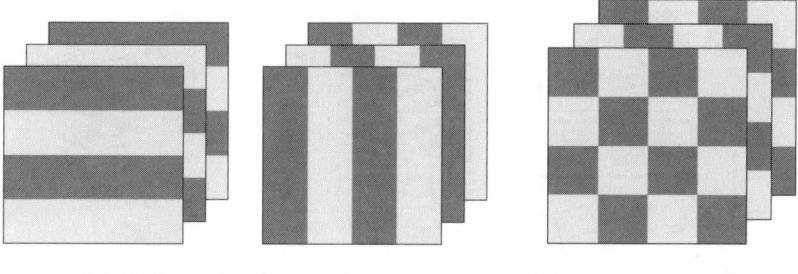

(a) Horizontal and vertical bars. (b) Checkerboard stimulus.

Fig. 3.12   Spatially non-uniform visual stimuli

The visual stimulation is generated using a 17" CRT high-resolution RGB monitor attached to a computer running a specifically designed program, written in the Python programming language using the VisionEgg libraries [Vision Egg (2007)]. The images are focused with the help of a lens onto the photoreceptor layer. To set up the whole system for data acquisition the retina is first flashed periodically with full field white light, whereas a microelectrode array (MEA), like the Utah microelectrode array (depicted in Fig. 3.13), is lowered into the retina until a significant number of electrodes detect light-evoked single and multi-unit responses. This allows the recording from 60-70 microelectrodes on average out of a total of 100 microelectrodes during each experiment. The electrode array is connected to a 100-channel amplifier (low and high corner frequencies of 250 and 7500 Hz) and to a digital signal processor-based data acquisition system. The neural spike events are detected by comparing the instantaneous electrode signal to the thresholds set for each data channel. When

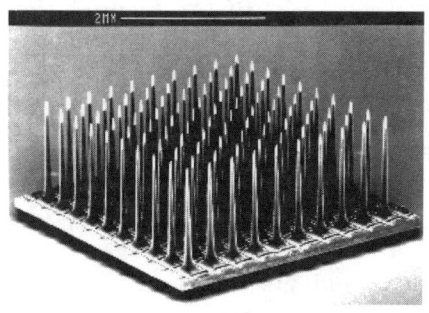

Fig. 3.13   The Utah Microelectrode Array.

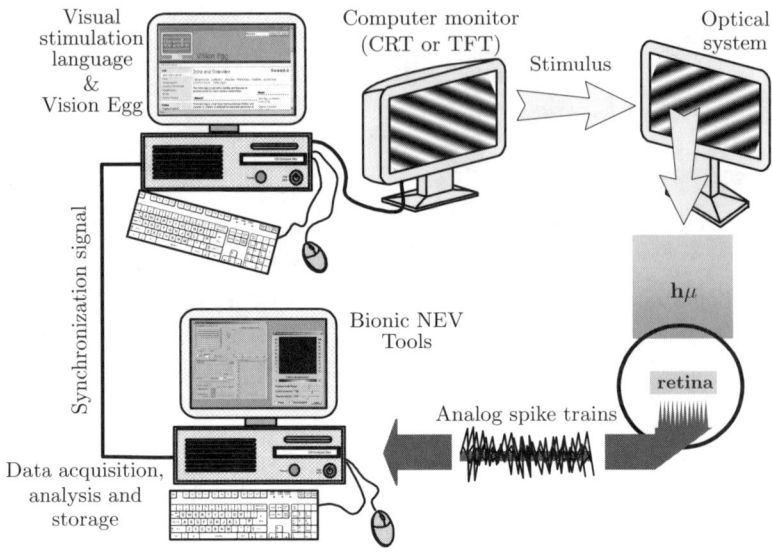

Fig. 3.14 Experimental apparatus for retina data acquisition and analysis.

a supra-threshold event occurs, the signal window surrounding the event is time-stamped (see Fig. 3.2), and is stored for later offline analysis. All the selected channels of data, as well as the state of the visual stimulus, are digitized with a commercial multiplexed A/D board data acquisition system, from Bionic Technologies, Inc, (now part of Cyberkinetics, Inc [Cyberkinetics (2008)]) and stored digitally.

The preparation of the retina for the data acquisition process needs to follow a delicate and sensitive process in order to obtain an effective and meaningful response from its RGCs. Figure 3.15 illustrates the whole process for the preparation of a rabbit retina. First, the animal is sacrificed with an injection of an overdose of anesthetic solution, followed by the enucleation of its eye. The eyeball is hemisected with a razor blade and the cornea and lens are separated from its posterior half. The retina is then carefully removed from the remaining eyecup with the pigment epithelium, and is mounted on a glass slide, with the ganglion cell layer side up, which is then covered with a millipore filter. This preparation is then mounted on a recording chamber and superfused with bicarbonate-buffered Ames solution at a temperature of 35° C, to postpone the retina death.

The recorded neuron cell responses are classified according to the spike

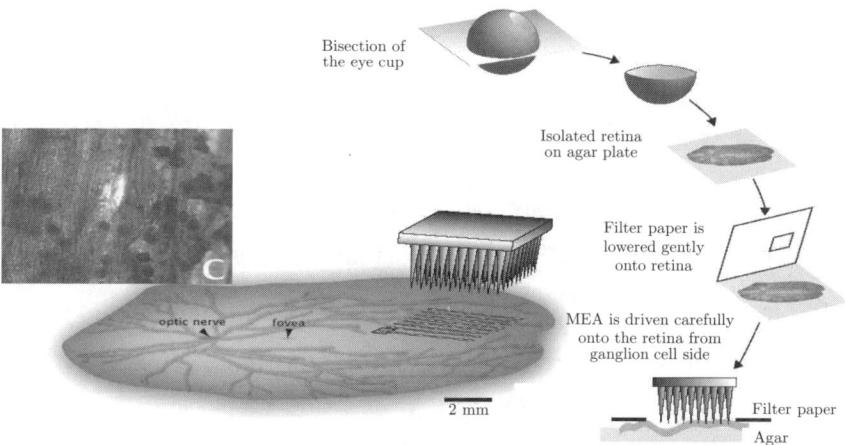

Fig. 3.15 Retina preparation for data acquisition.

waveform shapes, by means of the classification software SAC [Shoham (2001)], which uses a decomposition algorithm based on the Expectation-Maximization (EM) algorithm, in which the distribution of waveforms from each unit is modeled as a multivariate t-student distribution. As seen in Fig. 3.17, the neuron responses to the same stimulus are not unique; instead they a have a certain degree of randomness.

3.2.2 Spike Trains and Firing Rates

The spike train in Fig. 3.3 can be mathematically described as a series of Dirac delta functions, or unit impulse functions in the signal processing terminology [Oppenheim et al. (1999b)], each one positioned at a spike time occurrence t_i, where $i = 1, \ldots, n$, where n is the total number of spikes in the train:

$$\rho(t) = \sum_{i=1}^{n} \delta(t - t_i) , \qquad (3.7)$$

and $\rho(t)$ is the termed *neural response function*. The neural response function neglects the height and shape of the action potentials, so that all information is contained in the time arrival of the spikes; thus, the spike train is considered to be a point process [Brenner et al. (2002)].

The Dirac delta function $\delta(t)$ has a set of important properties [Arfken and Weber (2005)]:

$$\int_{-\infty}^{+\infty} \delta(t)dt = 1 , \qquad (3.8)$$

and

$$\delta(t) = 0 \quad \text{for} \quad t \neq 0 . \qquad (3.9)$$

Another useful property of the delta function, which is a direct consequence of the previous ones, is known as the sifting property:

$$\int_{-\infty}^{+\infty} \delta(t-t_0)f(t)dt = \int_{-\infty}^{+\infty} \delta(t-t_0)f(t_0)dt = f(t_0) . \qquad (3.10)$$

The Dirac delta function is not strictly a function since it does not have a closed mathematical definition. To define the continuous Dirac delta function, we can use the $\delta_\Delta(t)$ function depicted in Fig. 3.16 and defined as:

$$\delta_\Delta(t) = \begin{cases} \frac{1}{\Delta T} & \frac{-\Delta T}{2} < t < \frac{\Delta T}{2} \\ 0 & \text{otherwise} \end{cases} . \qquad (3.11)$$

By taking the limit when $\Delta T \to 0$ we have:

$$\delta(t) = \lim_{\Delta T \to 0} \delta_\Delta(t) . \qquad (3.12)$$

Given the definition in Eq. (3.12), we can see that the delta function has an area equal to one, and, by making $\Delta T \to 0$, the nonzero function values become concentrated around the origin, having an infinitesimal duration and an infinite amplitude. Although it has a peculiar definition, the enumerated properties make the delta function very useful.

The discrete-time counterpart of the delta function is the unit impulse. Unlike its continuous form, the discrete unit impulse has a closed mathematical formula (see [Oppenheim et al. (1999a)]):

$$\delta[n] = \begin{cases} 0 & n \neq 0 \\ 1 & n = 0 \end{cases} , \qquad (3.13)$$

where n in Eq. (3.13) represents the discrete independent variable, usually time[1].

---

[1]Due to the fact that n is frequently used for the independent variable in discrete-time, we decided to use it also here. However, n is also employed here to denote the total number of spikes in a trial. The two meanings should be distinguished according to the context.

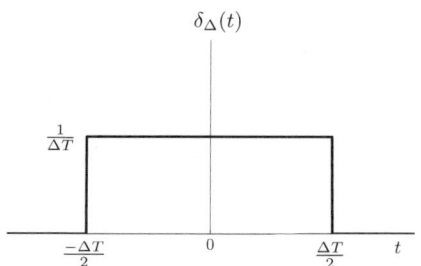

Fig. 3.16 The δ_Δ function.

From Eq. (3.7), we can obtain the number of spikes, n, present in the recording of a neuron response with a total time duration T and where each spike occurs at instants t_i, so that $0 \leq t_i \leq T$ for all i by integrating the neuron response:

$$\begin{aligned} n &= \int_0^T \rho(\tau)d\tau \\ &= \sum_{i=1}^n \int_0^T \delta(\tau - t_i)d\tau \;, \end{aligned} \qquad (3.14)$$

where the integral in the last term evaluates to one due to the property in Eq. (3.8).

The neural response can be characterized by several quantities computed from the neural function. One such quantity is the *spike-count rate*. The spike-count rate, r, is the number of spikes, n, appearing during a trial, divided by the total time duration, T, of the trial, and has the expression:

$$r = \frac{n}{T} = \frac{1}{T}\int_0^T \rho(\tau)d\tau \;. \qquad (3.15)$$

The spike-count rate is the time average of the neural response over a particular trial. The spike-count rate does not give any temporal information about the neural response.

We can average the neural response function over many experimental trials obtained with the same stimulus, such as the trials displayed in

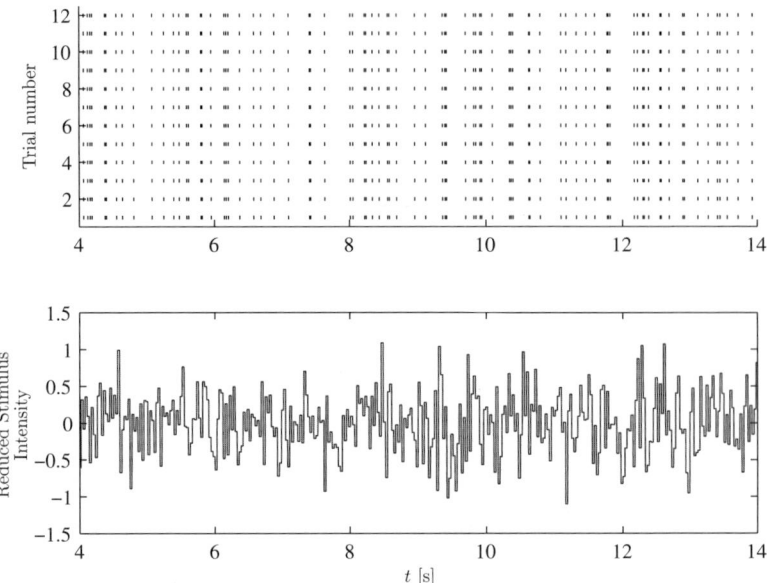

Fig. 3.17 Neural spike trains from a Salamander ON-type retinal ganglion cell (*Top*) when driven by the uniform white noise stimulus obtained from sampling a Gaussian distribution (*Bottom*), (data from [Keat et al. (2001)]).

Fig. 3.17. This average is denoted by $\langle \rho(t) \rangle$, and is computed by summing all the individual neural response functions for each trial and dividing the result by the number of trials M. If the neural response function for the trial j, where $j = 1, \cdots, M$, is represented by $\rho_j(t)$, the expression for the *average neural response* is:

$$\langle \rho(t) \rangle = \frac{1}{M} \sum_{j=1}^{M} \rho_j(t) \; . \tag{3.16}$$

If we represent the occurrence of a spike at time t_i, with $0 \le t_i \le T$, in the trial j, for $j = 1, \cdots M$, as t_{ij} then, with this nomenclature, the neural response average can be written as:

$$\langle \rho(t) \rangle = \frac{1}{M} \sum_{j=1}^{M} \sum_{i=1}^{n_j} \delta(t - t_{ij}) \; , \tag{3.17}$$

where n_j represents the total number of spikes in the trial j, which is usually different from trial to trial.

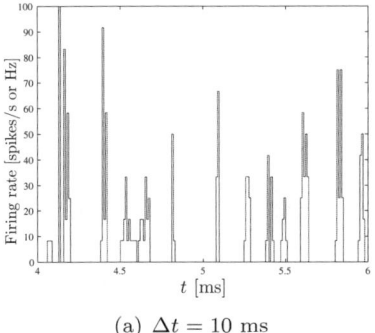
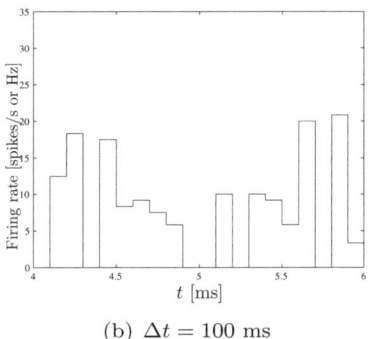

(a) $\Delta t = 10$ ms (b) $\Delta t = 100$ ms

Fig. 3.18 Retinal ganglion cell ON-type firing rate using different bin widths for the neuron response division.

A *time-dependent firing rate* can be obtained by counting spikes over shorts intervals of time, which is done more precisely with a large number of trials, which results from repeatedly presenting the same stimulus to the neuron. The firing rate at time t can be computed by counting the number of spikes that occur between the time instants t and $t + \Delta t$ and dividing the result by Δt; thus, the firing rate is measured in spikes/s or Hz. The precision of the firing rate computation increases by using a narrower time bin width Δt; as a result, a higher temporal resolution can be achieved. If we use only a spike train from a single trial and a narrow time bin, the firing rate will be either zero or one, so an average over multiple trials should be made. The time-dependent firing rate is defined as the average number of spikes over several trials appearing in an interval between time instants t and $t + \Delta t$, divided by the interval length Δt. The time-dependent firing rate is defined as:

$$r(t) = \frac{1}{M} \sum_{j=1}^{M} \frac{1}{\Delta t} \int_{t}^{t+\Delta t} \rho_j(\tau) d\tau$$

$$= \frac{1}{\Delta t} \int_{t}^{t+\Delta t} \langle \rho(\tau) \rangle d\tau ,$$

(3.18)

where the integral in the first equality computes the number of spikes between the times t and $t+\Delta t$ in the jth trial, while the integral in the second equality captures the number of spikes in the time range t to $t + \Delta t$ of the neural response average.

Formally, we can make the time interval length of Eq. (3.18) go to zero. Thus, by taking limit we have

$$r(t) = \lim_{\Delta t \to 0} \frac{1}{\Delta t} \int_{t}^{t+\Delta t} \langle \rho(\tau) \rangle d\tau$$
$$= \left\langle \lim_{\Delta t \to 0} \frac{1}{\Delta t} \int_{t}^{t+\Delta t} \rho(\tau) d\tau \right\rangle, \quad (3.19)$$

where we can recognize the derivative of the integral of $\rho(t)$ inside the angle brackets in the last expression. Therefore, the formal definition of the firing rate can be written as:

$$r(t) = \langle \rho(t) \rangle, \quad (3.20)$$

that states that the firing rate is the average of the neural function over many trials.

To calculate the firing rate from the observed data, a time interval with a finite length must be used in order to obtain a reliable estimate of the average. Figure 3.18 displays a parcel of the firing rate for the data in Fig. 3.17. For a small Δt, the quantity $r(t)\Delta t$ is the average number of spikes present at the interval $[t, t + \Delta t]$:

$$\langle n \rangle_{[t,t+\Delta t]} = \int_{t}^{t+\Delta t} r(t) dt \quad (3.21)$$
$$\cong r(t)\Delta t.$$

By integrating $r(t)$ over a given time interval, we obtain the average number of spikes over that interval.

If Δt is small in Eq. (3.18), there will be no more than one spike in the time interval $t \in [t, t + \Delta t]$ for each neural response function, so that $r(t)\Delta t$ gives the fraction of trials in which a spike occurred between those times. In this way, $r(t)\Delta t$ can be interpreted as the probability of a spike occurring in the interval Δt around t. That is,

$$P(n = 1 \text{ in } \Delta t) = r(t)\Delta t. \quad (3.22)$$

The spike-count firing rate can be averaged over several trials, yielding a another quantity called the *average firing rate* and denoted by $\langle r \rangle$ with the expression

$$\langle r \rangle = \frac{\langle n \rangle}{T} = \frac{1}{T} \int_{0}^{T} \langle \rho(\tau) \rangle d\tau = \frac{1}{T} \int_{0}^{T} r(\tau) d\tau. \quad (3.23)$$

The first equality in Eq. (3.23) indicates that $\langle r \rangle$ is the average number of spikes per trial divided by the trial duration, where the second and third equalities result from Eq. (3.14) and Eq. (3.20), respectively.

The three different quantities: r, $r(t)$, and $\langle r \rangle$, are frequently used in the literature without distinction as *firing rate*, which causes some misunderstanding; therefore, special care should be taken to understand from the context which definition is being employed.

The formal firing rate $r(t)$ defined by Eq. (3.19) should, theoretically, be calculated with an infinite set of trials. This is impossible, but the estimate becomes more accurate as more and more trials are included in the set. Indeed, we have only a finite set of experimental trials that must be discretized into time bins, and the firing rate can be obtained by counting spikes within the time bins that compose a trial and by averaging over trials.

In practice, the firing rate is estimated by dividing the spike trains into time bins with a finite width Δt, then the number of spikes within each time bin is counted, and the result divided by Δt. Furthermore, if we have several trials as the response to the same stimulus, we can perform this operation for all trials and average the results over these trials. This procedure generates a staircase-like spike-count firing rate, which is a piecewise constant time function for each time bin, resembling a histogram (see Fig. 3.18). Each bar amplitude is equal to $1/\Delta t$ times the number of spikes in each time bin. By decreasing the width of the time bins, Δt, the temporal resolution increases, and the firing rate can be estimated at a finer time scale, but at the cost of reducing the number of possible different rates. In the limiting case, for a time bin narrower than the neuron's refractory period, there can be only 0 or 1 spike in each bin, so that the height of each histogram bar can be only 0 or $1/\Delta t$.

One way to avoid the quantization of the firing rate, which is always proportional to the inverse of the time bin width, is to divide the trial into bins with a variable width by fixing the number of spikes within each time bin, so that the firing rate is approximated by the fixed number of spikes in the bin divided by the variable bin width.

Both procedures described make the firing rate dependent on the size of the bin and/or the spikes' locations. To prevent these effects a time window, $w(t)$, can be used to smooth the firing rate. This window slides along the spike trains and counts the number of spikes within the window at each new location. This is the same as convolving the spike trains with the window. Generalizing, if we have M trials with spikes positioned at times t_{ij}, with $i = 1, \cdots, n_j$, where n_j is the number of spikes in the trial j, and $j = 1, \cdots, M$, the firing rate can be approximated by

$$r_{\text{approx}}(t) = \int_{-\infty}^{+\infty} w(\tau) \langle \rho(t-\tau) \rangle d\tau$$

$$= \frac{1}{M} \sum_{j=1}^{M} \sum_{i=1}^{n_j} w(t - t_{ij}), \quad (3.24)$$

for a window with unit area (otherwise the result must be divided by the window area). This procedure corresponds to the convolution of the average neural response with the filter kernel $w(t)$. The last equality in Eq. (3.24) results from the Dirac delta property in Eq. (3.10).

Convolution is one of the most important operations in signal processing and system analysis since many systems can be completely specified by their convolution properties. The convolution operator is denoted by an asterisk, and it represents the integral in continuous time:

$$y(t) = x(t) * h(t) = \int_{-\infty}^{+\infty} x(\tau) h(t-\tau) d\tau . \quad (3.25)$$

The correspondent operation in discrete-time is defined as [Oppenheim et al. (1999b)]:

$$y[n] = x[n] * h[n] = \sum_{k=-\infty}^{+\infty} x[k] h[n-k] . \quad (3.26)$$

Using this notation, the approximation for the firing rate in Eq. (3.24) can be written as:

$$r_{\text{approx}}(t) = w(t) * \langle \rho(t) \rangle . \quad (3.27)$$

One of the simpler filter kernels is the rectangular window, or boxcar filter. A rectangular window with duration ΔT and unit area has the expression in continuous time:

$$w_{\text{rect}}(t) = \begin{cases} \frac{1}{\Delta T} & -\frac{\Delta T}{2} < t < \frac{\Delta T}{2} \\ 0 & \text{otherwise} \end{cases} . \quad (3.28)$$

Figure 3.19 graphically represents a rectangular window. As a consequence of using a window to convolve the spike trains, the values obtained for the firing rates separated in time by less than one window width are correlated since they include common spikes in their calculation.

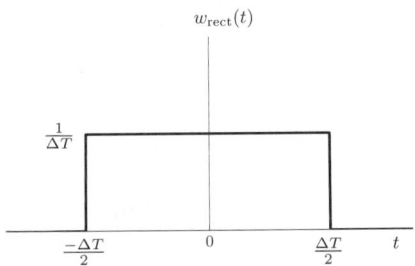

Fig. 3.19   The rectangular (boxcar) filter window.

The main problem with the rectangular window is that it is not a continuous function. As a result, filtering the neural function average, which is also a series of discontinuous delta functions, results in a irregular waveform for the firing rate. To avoid these effects, a continuous window function that goes to zero as the distance to the origin increases can be used. A widespread filter window is the ubiquitous bell-shaped Gaussian function. The Gaussian function is defined as

$$w_{\text{gauss}}(t) = \frac{1}{\sqrt{2\pi}\sigma_w} e^{-\frac{t^2}{2\sigma_w^2}} , \qquad (3.29)$$

where the parameter σ_w controls the width of the function and, consequently, the temporal resolution of the rate. Figure 3.20 displays the graph of the Gaussian window.

Both the rectangular and Gaussian windows are not causal, meaning that they approximate the firing rate at a given time instant by taking into account spikes that were fired before and after that instant. To make the firing rate at time t dependent only on spikes fired before t, we must use a causal filter for the window. A commonly used causal filter is the α window (see [Dayan and Abbot (2001)]), which is expressed as

$$w_\alpha(t) = \alpha^2 t \exp(-\alpha t)\, \text{u}(t) , \qquad (3.30)$$

where $1/\alpha$ controls the temporal resolution of the firing rate estimate. The α function is displayed in Fig. 3.21. The expression of Eq. (3.30) introduces the continuous Heaviside function, u(t), or the *unit step function*, in signal processing terminology [Ziemer et al. (1998)], defined as:

$$u(t) = \begin{cases} 0 & t < 0 \\ 1 & t \geq 0 \end{cases}. \qquad (3.31)$$

The Dirac delta function, defined in Eq. (3.12) as the limit of a rectangular window, can also be defined as the derivative of the Heaviside unit step function:

$$\delta(t) = \frac{d\,u(t)}{dt}. \qquad (3.32)$$

The delta function properties, introduced in Eq. (3.8) and Eq. (3.9), still hold. The delta function is called a generalized function that can be defined as the limit of several alternative functions as long as its properties are satisfied [Ziemer et al. (1998); Arfken and Weber (2005)]. The discrete-time counterpart of the continuous Heaviside unit step function, Eq. (3.31), is the *unit step sequence* with the expression:

$$u[n] = \begin{cases} 1 & n \geq 0 \\ 0 & n < 0 \end{cases}, \qquad (3.33)$$

where n is the discrete independent variable. The unit step sequence is related with the unit impulse, defined in Eq. (3.13), by

$$\delta[n] = u[n] - u[n-1] \quad \text{or} \quad u[n] = \sum_{k=-\infty}^{n} \delta[k]. \qquad (3.34)$$

In order to process the neuronal data in a computer, the data must be discretized. In discrete-time, the spikes in the continuous neural function

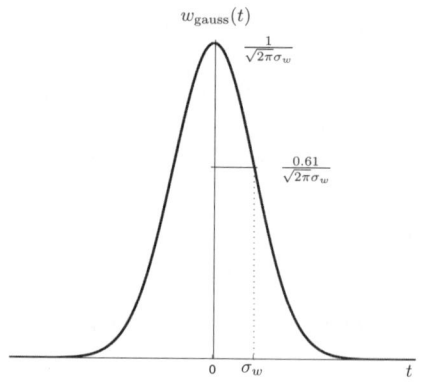

Fig. 3.20 The Gaussian filter window.

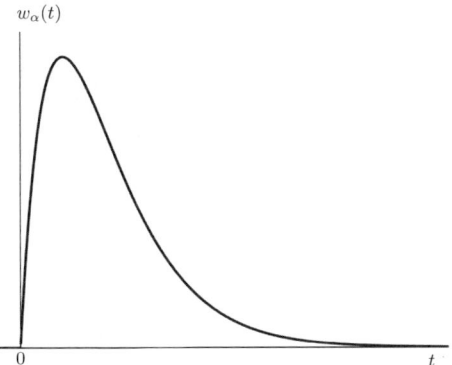

Fig. 3.21 The α function filter.

of Eq. (3.7) cannot be positioned at any arbitrary instant of time. In the process of discretization, if we use a sampling period Δt, the spikes are placed within a specific time bin, n_i, with width Δt. In discrete-time, the neuronal function of Eq. (3.7) is represented by a series of discrete delta functions, or unit impulses, defined in Eq. (3.13), as:

$$\rho[n] = \sum_{i=1}^{n} \delta[n - n_i], \quad (3.35)$$

where n_i are the discrete time instants corresponding to the time bins containing a spike. The discrete neural function can be obtained by dividing the total time duration of the response, T, into intervals of width Δt, so that the sequence of the neural function has a length equal to $N = T/\Delta t$. It is convenient to make Δt small, which means that the sampling frequency has to be sufficiently high in order to have at most one spike in each time bin, which can be accomplished by choosing a time sampling period smaller than the refractory period of the neuron. It was shown that a sampling frequency within the range 10 kHz to 20 kHz is enough for the case of primate retinal ganglion cells [Uzzell and Chichilnisky (2004)].

The firing rate, defined in Eq. (3.20) for the continuous case, becomes in discrete-time

$$r[n] = \frac{1}{\Delta t} \frac{1}{M} \sum_{j=1}^{M} \sum_{i=1}^{n_j} \delta[n - n_{ij}], \quad (3.36)$$

where the limit n_j in the sum represents the number of spikes in the trial j and the limit M is the total number of trials in the experiment. The

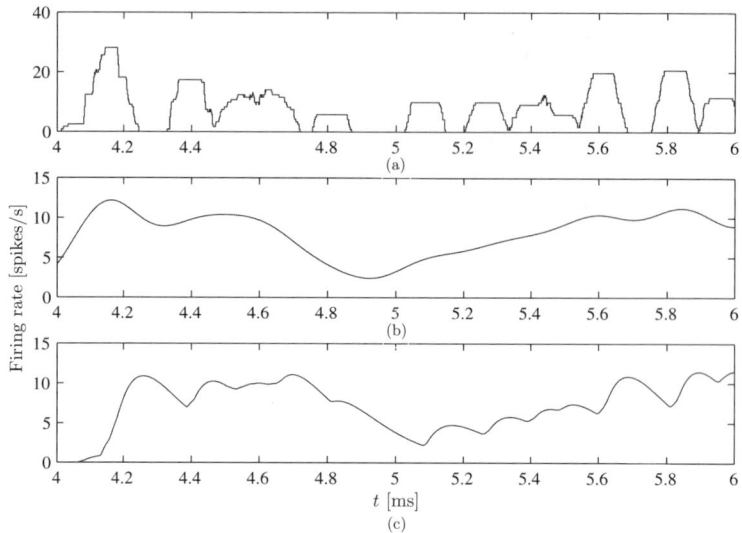

Fig. 3.22 Firing rate obtained by filtering the neural response with different types of filter windows: (a) rectangular window with $\Delta t = 100$ ms; (b) Gaussian window with $\Delta t = 100$ ms; (c) α-window with $1/\alpha = 100$ ms. The sampling rate is $F_s = 1$ kHz.

resulting sequence $r[n]$ gives the number of fired spikes per second (Hz) in the time bin n, with $0 \leq n \leq N-1$. The discrete firing rate $r[n]$ is frequently plotted against time, giving the peri-stimulus time histogram (PSTH), as depicted in Fig. 3.22.

The average firing rate is obtained from the discrete firing rate by the expression:

$$\langle r \rangle = \frac{1}{N} \sum_{n=1}^{N} r[n] , \qquad (3.37)$$

where N is the number of time bins into which the spike train was sampled. The average number of spikes $\langle n \rangle$ within the time bin n, corresponding to the time interval $[n\Delta t, (n+1)\Delta t]$, is given by

$$\langle n \rangle_{[n\Delta t, (n+1)\Delta t]} = r[n]\Delta t . \qquad (3.38)$$

For a small Δt, so that there is at most one single spike per time bin, the firing rate corresponds to the probability that the neuron will fire a spike within that time bin.

3.2.3 Spike Triggered Average

A useful and enlightening characteristic of a neuron is the waveform of the stimuli that produces a given response. We can choose different patterns of spikes and compute the stimuli that originate it. There are several possibilities, ranging from the stimulus that elicits a single spike to the stimulus that produces a more complicated sequence of spikes [de Ruyter van Steveninck and Bialek (1988)]. However, the most simple and common situation is to analyze the stimuli that elicited a single spike.

With the single spike response, we try to figure out what the stimulus looks like, on average, before this single action potential was fired. The resulting quantity is termed the spike triggered average (STA). The STA provides a very useful technique to characterize the neural selectivity, and it constitutes the basic framework for several retina models [Chichilnisky (2001); de Ruyter van Steveninck and Bialek (1988)]. The STA is also referred to in the literature as the reverse correlation function, the mean effective stimulus, the triggered correlation function, or even the first Wiener kernel [Rieke et al. (1997)].

The procedure to compute the STA is: *i*) pick the stimulus segments before every fired spike in a trial, *ii*) add all these segments, *iii*) normalize the resulting waveform by the total number of spikes, and then *iv*) repeat the previous procedure for all trials available and normalize the result by the number of trials. Figure 3.23 illustrates the computing process of the STA for a single spike train #j with n_j spikes.

If we represent the stimulus segment with length $\tau_{\max}$, which occurs just before a spike located at the time instant t_i in trial j by $s(t_{ij} - \tau)$, with $0 < \tau \leq \tau_{\max}$ for $i = 1, \cdots, n_j$, where n_j is the number of spikes in trial j, and $j = 1, \cdots, M$, where M is the number of trials evolved in the computation, the STA takes the value

$$s_{\text{spk}}(\tau) = \frac{1}{M} \sum_{j=1}^{M} \frac{1}{n_j} \sum_{i=1}^{n_j} s(t_{ij} - \tau) \qquad 0 \leq \tau \leq \tau_{\max} . \qquad (3.39)$$

This expression is equivalent to a weighted sum over several trials.

Although it was not imposed a limit for the value of $\tau_{\max}$ in Eq. (3.39), the neural response depends on the stimulus only within a time window with a few hundred milliseconds wide before the spike occurrence, which corresponds to the neuron memory. This happens because $s_{\text{spk}}(\tau)$ goes to zero for positive values of τ larger than the correlation time between the stimulus and the response. If, however, the stimulus is not temporally

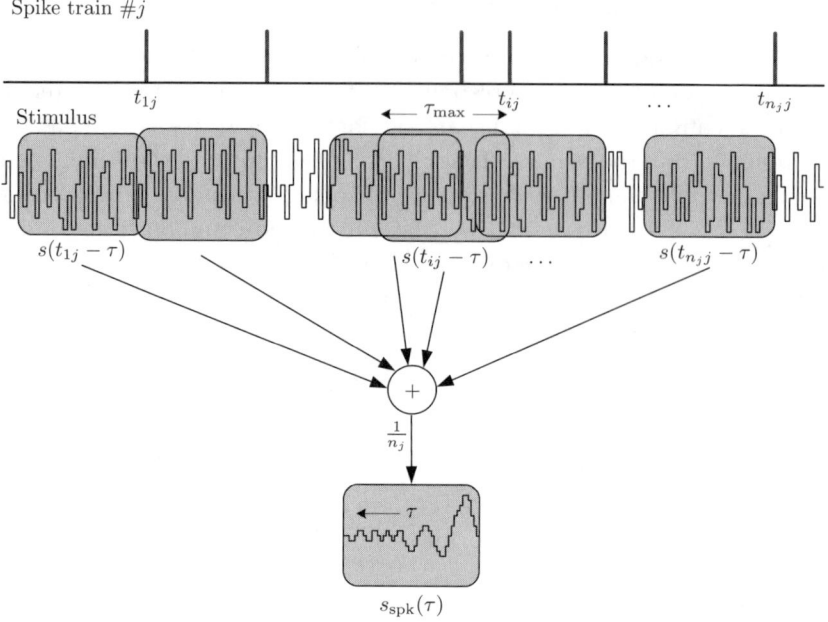

Fig. 3.23 Procedure for the STA computation.

autocorrelated, we can state that $s_{\text{spk}}(\tau)$ will be 0 for $\tau < 0$ because the neuron response should not be dependent on future stimuli – it is a causal system.

To compute the STA, the stimuli segment is acquired over a finite time period before every spike occurrence; these different stimuli are summed, and the result is normalized by the number of spikes considered. Figure 3.24 shows the plot of the STA, time reversed, for two distinct types of RGC cells. These cells are categorized according to the waveform of their STA. As we can see from Fig. 3.24(a), the RGC of the salamander fires preferably in response to the positive onset of stimulus, so it is classified as an ON-type cell. The rabbit RGC in Fig. 3.24(b) fires predominantly when the stimulus has a negative offset, so it is classified as an OFF-type cell.

Taking into account the property expressed in Eq. (3.10) we can write

$$s(t_i - \tau) = \int_0^T \delta(t - t_i) s(t - \tau) dt \ , \tag{3.40}$$

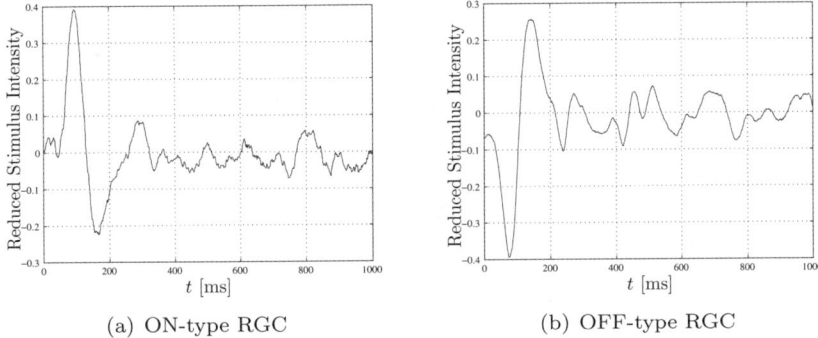

Fig. 3.24 Spike triggered average (STA), time reversed, of a ON-type retinal ganglion cell (salamander) and of an OFF-type cell (rabbit), (data from [Keat et al. (2001)]).

so that Eq. (3.39) becomes:

$$s_{\text{spk}}(\tau) = \frac{1}{M} \sum_{j=1}^{M} \frac{1}{n_j} \sum_{i=1}^{n_j} \int_0^T \delta(t - t_{ij}) s(t - \tau) dt$$

$$= \frac{1}{M} \sum_{j=1}^{M} \int_0^T \frac{1}{n_j} \sum_{i=1}^{n_j} \delta(t - t_{ij}) s(t - \tau) dt$$

(3.41)

By assuming that each spike train has a similar number of spikes, we can approximate the number of spikes in each trial by $n_j \simeq \langle n \rangle$; for example, this is a good approximation when the number of spikes in each trial is high. With this assumption, Eq. (3.41) can be written as:

$$s_{\text{spk}}(\tau) = \frac{1}{\langle n \rangle} \left\langle \int_0^T \sum_{i=1}^{n} \delta(t - t_i) s(t - \tau) dt \right\rangle$$

$$= \frac{1}{\langle n \rangle} \left\langle \int_0^T \rho(t) s(t - \tau) dt \right\rangle$$

(3.42)

The last equation above relates the STA with the neural response function defined by Eq. (3.7). Furthermore, if the same stimulus is used for all trials, Eq. (3.42) can be further simplified to:

$$s_{\text{spk}}(\tau) = \frac{1}{\langle n \rangle} \int_0^T \langle \rho(t) \rangle s(t - \tau) dt = \frac{1}{\langle n \rangle} \int_0^T r(t) s(t - \tau) dt \ . \quad (3.43)$$

By definition, the correlation function of the firing rate with the stimulus is:

$$C_{rs}(\tau) = \frac{1}{T}\int_0^T r(t)s(t+\tau)dt \;. \tag{3.44}$$

Equation (3.43) relates the STA with the correlation function of the stimulus with the firing rate. Comparing the previous equation with Eq. (3.43), we have the relation

$$s_{\text{spk}}(\tau) = \frac{1}{\langle r \rangle} C_{rs}(-\tau)\;, \tag{3.45}$$

where $\langle r \rangle = \langle n \rangle / T$ is the average firing rate over the set of trials. From Eq. (3.45), we can see that $s_{\text{spk}}(\tau)$ is proportional to the correlation function of the firing rate with the stimulus at preceding times, so that the STA is also called the *reverse correlation function*.

Whenever the input stimulus has a shape similar to the STA, the neuron has a high probability of firing a spike. This is the genesis of several neuron models, and of retina models in particular. Considering that the firing rate of a visual neuron is a functional of a filtered version of the input stimulus $s(t)$,

$$r(t) = r_0\, F\left[\int_{-\infty}^{\infty} h(\tau)s(t-\tau)d\tau\right], \tag{3.46}$$

where $h(t)$ is the filter's impulse response and $F[\cdot]$ is a nonlinear memoryless functional. It has been shown that the STA is proportional to the filter $h(t)$ [Rieke et al. (1997); Chichilnisky (2001)]. This result is very useful since it allows to separate the calculus of the linear filtering properties of the neuron response from the nonlinearities in the spike generation [Keat et al. (2001)]. The STA can also be interpreted as the neuron receptive field since it represents the preferred stimulus of the ganglion cell in terms of time profile [Rust et al. (2004)].

We must be aware that, in the expression for the STA of Eq. (3.39), the stimulus segments gathered from different trials are weighted differently according to the number of spikes contained in the trial – the stimulus segments belonging to a train with fewer spikes have a greater weight in the overall sum. However, the important relations between Eq. (3.39) and other quantities, such as the neural function in Eq. (3.42), the firing rate in Eq. (3.43), and the correlation function in Eq. (3.45), are valid only if the approximation $n_j \simeq \langle n \rangle$ is fulfilled.

We can simplify the calculation of the STA for discrete-time. If the sequence $s[n]$ represents the discrete input stimulus of a RGC, whose response is the binary sequence $\rho[n]$, which is composed by a sequence of unit impulse functions, then we can calculate the STA using the expression

$$s_{\text{spk}}[l] = \frac{\sum_{j=1}^{M}\sum_{n=1}^{N} \rho_j[n]s[n-l]}{\sum_{j=1}^{M}\sum_{n=1}^{N} \rho_j[n]}, \qquad 0 \le l \le l_{\max}, \qquad (3.47)$$

where N is the length of the stimulus sequence and $\rho_j[n]$ is the neural response of the trial j of a total of M responses obtained in the experiment by applying the same stimulus. In Eq. (3.47), all stimulus segments before every spike in every available trial are summed, and the result is divided by the total number of spikes occurring in all trials. In matrix notation, the STA can be represented by the vector $\mathbf{s}_{\text{spk}}$, with dimension $l_{\max}$ corresponding to the neuron memory.

The STA is particularly meaningful if the stimulus is composed of a white noise sequence, meaning that it is uncorrelated. For a stimulus sequence that is not autocorrelated, the resulting STA sequence goes to zero for a time lag larger than the neuron memory. The time lag, $l_{\max}$, is the discrete dual of τ_{max} in Eq. (3.39) for the continuous, and they are related by $l_{\max} = \tau_{\max}/\Delta t$, where Δt is the sampling period used in the discretization of the stimulus and of the neural response. Depending on the species, a time lag between 40 ms and 950 ms is used [Keat et al. (2001)] to calculate the STA of the retinal ganglion cells.

Although we have considered only time in showing how to calculate the STA, we can also include space. To calculate the spatial spike-triggered average, we must average the sequence of images (and not only a single luminance value for each time instant) that generated a fire for every spike occurrence. This is the procedure to obtain, for example, the spatial and temporal form of the receptive field of a neuron [Chichilnisky (2001)].

3.2.4 Spike Train Autocorrelation Function

The *spike train autocorrelation function* gives the time relation between any two spikes in a given spike train. It gives the distribution of times between any two spikes in a train.

The spike train autocorrelation function is the autocorrelation of the quantity obtained by subtracting from the neural function, defined in

Eq. (3.7), the mean firing rate $\langle r \rangle$ averaged over time and over trials:

$$C_{\rho\rho}(\tau) = \frac{1}{T} \int_0^T \langle (\rho(t) - \langle r \rangle)(\rho(t+\tau) - \langle r \rangle) \rangle dt \ . \tag{3.48}$$

In fact, the expression in Eq. (3.48) follows the autocovariance definition since the mean firing rate is subtracted from the neural function before the autocorrelation computation [Therrien (1992)]. Nonetheless, it is called autocorrelation in the neuroscience literature [Dayan and Abbot (2001)].

We can define, analogously to the spike train autocorrelation, a cross-correlation between spike trains generated by different neurons. The *cross-correlation function* between a pair of spike trains is useful when we are looking for synchronicity in the firing from different neurons. For example, if the cross-correlation between two spike trains from different neurons have a peak at zero lag means that the neurons are firing synchronously, while a shift of the peak from zero indicates that the neurons are firing synchronously with a phase shift. The autocorrelation function is an even function of the lag τ so that $C_{\rho\rho}(\tau) = C_{\rho\rho}(-\tau)$, while the cross-correlation function is not an even function of the lag, which means that $C_{\rho_1\rho_2}(\tau) \neq C_{\rho_1\rho_2}(-\tau)$.

In practice, the spike train autocorrelation is computed by: *i)* sampling the continuous neural function into time bins of width Δt; *ii)* the number of spikes in each time bin is recorded into a sequence; *iii)* the spike count sequence, corresponding to the mean, is subtracted from this sequence; finally, *iv)* the autocorrelation of the sequence is computed. The resulting autocorrelation can be plotted in the form of a histogram as a function of the lag between the time bins.

In discrete-time, the spike train autocorrelation function is calculated by the autocorrelation of the neural response with the average firing rate removed, as shown by the expression:

$$C_{\rho\rho}[l] = \frac{1}{N} \frac{1}{M} \sum_{j=1}^{M} \sum_{n=1}^{N} (\rho_j[n] - \langle r \rangle)(\rho_j[n+l] - \langle r \rangle) \ , \tag{3.49}$$

where l represents the lag between two spikes measured in number of time bins. As before, N is the total number of time bins, corresponding to the neural response length, and M is the number of trials included in the calculation.

3.2.5 The Spike Triggered Covariance

The *spike-triggered covariance* is another important quantity that characterizes the neural response. Its application is gaining increasing relevance in recent studies that examine the modeling of neuronal responses, particularly of the retinal response [Schwartz et al. (2002); Simoncelli et al. (2004)]. The spike triggered covariance (STC) is used to obtain parameters from the firing rate second order statistics, and it can also be used to recover a series of linear filters that resemble the neural response in the presence of both symmetric and asymmetric nonlinearities [Rust et al. (2004)].

Qualitatively, the STC function tells us how the stimulus varies with itself before a spike is fired. It is obtained by computing the covariance between the stimulus segments before every spike. If the occurrence of spike i in the trial j is represented by t_{ij}, where $1 \leq i \leq n_j$ and $1 \leq j \leq M$, the STC is defined in continuous time by

$$C_{ss}(\tau_1, \tau_2) = \frac{1}{\sum_{j=1}^{M} \int_0^T \rho_j(t) dt}$$

$$\times \sum_{j=1}^{M} \sum_{i=1}^{n_j} [(s(t_{ij} - \tau_1) - s_{\text{spk}}(\tau_1)).(s(t_{ij} - \tau_2) - s_{\text{spk}}(\tau_2))] \quad (3.50)$$

$$= \frac{1}{M \sum_{j=1}^{M} n_j} \sum_{j=1}^{M} \sum_{i=1}^{n_j} s(t_{ij} - \tau_1).s(t_{ij} - \tau_2) - s_{\text{spk}}(\tau_1) s_{\text{spk}}(\tau_2) ,$$

where $s_{\text{spk}}(\tau)$ is the spike triggered average and the denominator corresponds to the total number of spikes from all trials. The STC is a bi-dimensional function that returns the stimulus variation at time τ_1, before the spike is fired, as a function of its value at time τ_2, before the spike is fired. In discrete-time, the expression for the STC is

$$C_{ss}[l_1, l_2] = \frac{1}{M \sum_{j=1}^{M} n_j} \sum_{j=1}^{M} \sum_{i=1}^{n_j} [s[n_{ij} - l_1] - s_{\text{spk}}[l_1]).(s[n_{ij} - l_2] - s_{\text{spk}}[l_2])] ,$$

(3.51)

where n_{ij} is the bin of the trial j where the spike i occurs and $s_{\text{spk}}[l]$ is the discrete spike-triggered average.

If the stimulus vector before spike i in trial j is represented by $\mathbf{s}_{ij}$,

$$\mathbf{s}_{ij} = \begin{bmatrix} s[n_{ij}] \\ s[n_{ij}-1] \\ \vdots \\ s[n_{ij} - l_{\max}] \end{bmatrix}, \qquad (3.52)$$

the STC can be written as:

$$\mathbf{C}_{\text{spk}} = \frac{1}{\sum\limits_{j=1}^{M} n_j} \sum_{j=1}^{M} \sum_{i=1}^{n_j} (\mathbf{s}_{ij} - \mathbf{s}_{\text{spk}})(\mathbf{s}_{ij} - \mathbf{s}_{\text{spk}})^T, \qquad (3.53)$$

where $\mathbf{s}_{\text{spk}}$ is the STA vector.

3.3 Stimulus and Response Statistics, and Firing Probabilities

As stated before, to estimate a probability with some reliability, we need a large number of data vectors. Or, if we are interested only in some common statistical parameters, such as the mean and variance, we need a very long experiment in time for the case of an ergodic process. The former case is more common in experimental neuroscience: the same stimulus is repeatedly presented to the neuron, and the respective responses are recorded. For the case of the retina neural circuit, the responses to the same stimulus show some variability that justifies, and even enforces, the application of probability tools to characterize and describe the neural code.

In a common experimental setup, the retina is excited with a visual stimulus, $s(t)$, chosen by the experimenter. As a consequence, a sequence of spikes, occurring at times $t_1, t_2, \cdots, t_n$, which correspond to a particular neural response, $\rho_j(t)$, is recorded. Then, the probability that the neuron will fire conditioned to that particular stimulus, $P\left(\rho_j(t)|s(t)\right)$, can be estimated from the collected responses.

The stimulus itself can be drawn from a given probability distribution $P[s(t)]$, such as a Gaussian distribution, for example, defining an ensemble of stimuli signals. The stimulus distribution may even resemble the statistical properties of natural scenes [Yu and de Sa (2004)].

Since the stimulus is random and the neuron response has also a random nature, the neuron activity can be described by the joint probability distribution of the stimuli signals and the spike trains: $P\left(\rho_j(t), s(t)\right)$. This joint distribution quantifies the likelihood that, in the course of the experiment, the stimulus $s(t)$ and the spike train $\rho_j(t)$ will both be observed. By employing the relationship between the joint, the marginal, and the conditional probability distributions, the joint distribution can be written using the response distribution conditioned to a given stimulus as

$$P(\rho_j(t), s(t)) = P(\rho_j(t)|s(t))\ P(s(t))\ , \qquad (3.54)$$

where $P(s(t))$ is the stimulus marginal probability distribution. We can also try to find which spike train a given stimulus will trigger. As mentioned before, there is not a one-to-one relationship between a given stimulus and the generated spike train, but we can write the following joint probability distribution:

$$P(s(t), \rho_j(t)) = P(s(t)|\rho_j(t))\, P(\rho_j(t))\ . \qquad (3.55)$$

By equating the probability distributions of Eq. (3.54) and Eq. (3.55), we have

$$P(\rho_j(t)|s(t))\, P(s(t)) = P(s(t)|\rho_j(t))\, P(\rho_j(t))\ . \qquad (3.56)$$

From Eq. (3.56), we can follow two different perspectives about the neural code. From a modeling, or encoding, perspective we would like to know the distribution $P(\rho_j(t)|s(t))$ so that, given a certain stimulus, the most likely neural response can be obtained. We can then generate the spike train to stimulate the superior parts of the nervous system. The desired distribution has the expression

$$P(\rho_j(t)|s(t)) = \frac{P(s(t)|\rho_j(t))\, P(\rho_j(t))}{P(s(t))}\ . \qquad (3.57)$$

This last equality expresses Bayes' rule that relates marginal and conditional probabilities.

By taking the brain decoding perspective, which perceives the stimulus from the received spike train, the distribution of interest is:

$$P(s(t)|\rho_j(t)) = \frac{P(\rho_j(t)|s(t))\, P(s(t))}{P(\rho_j(t))}\ . \qquad (3.58)$$

Next, we will present the most common probability distributions employed in the analysis of the neural response.

3.3.1 Spike Train Statistics

The structure of a spike train can be perfectly regular, as in the trains generated by neurons controlling the heart beat where spikes are fired regularly at almost constant time intervals. On the opposite extreme, they can show a completely random behavior, where spikes are fired independently of past history, as is observed in the brain.

The retinal ganglion cells fire spontaneously, even in the dark, such that those spike trains obviously do not convey any visual information and should be considered as pure noise. However, when properly stimulated, retinal ganglion cells can produce very reproducible spike trains from trial to trial [Berry et al. (1997b)], which means that the spike train variability is controlled.

The stochastic process that generates a sequence of stereotyped discrete events, such as spikes, is called a *point process*. The probability $P(t_1, t_2, \ldots, t_n)$ that a sequence of n spikes occurring at the time instants $\{t_1, t_2, \ldots, t_n\}$ is proportional to the probability density of the spike occurrences at those times: $p(t_1, t_2, \ldots, t_n)$. Specifically, the probability of occurrence of a sequence of n spikes, with spike i occurring between time instants t_i and $t_i + \Delta T$, with $i = 1, 2, \ldots, n$, is given by:

$$P(t_1, t_2, \ldots, t_n) = p(t_1, t_2, \ldots, t_n) \cdot (\Delta T)^n . \qquad (3.59)$$

However, the possible number of different spike sequences is so huge that we would need an infinite amount of data to fully characterize the probability density $p(t_1, t_2, \ldots, t_n)$. The impossibility of reliably characterizing the probability density in Eq. (3.59) to a desirable degree led to the development of statistical models that simplify the description of a spike train.

In general, the generation of a spike can depend on the full history of the process. That is, the generation of a spike at time t_n can depend on all the spikes generated previously, so that the probability of a neuron firing a spike at the time instant t_n is dependent on all the previously generated spikes. This is described mathematically as the conditional probability

$$P(t_n | t_1, t_2, \ldots, t_{n-1}) . \qquad (3.60)$$

By observing the firing dynamics of a neuron, it was noticed that, for most cases, the firing of a spike depends mainly on the last generated spike. If the probability of spike generation depends only on the last generated spike, the conditional probability of Eq. (3.60) can be simplified to

$$P(t_n | t_1, t_2, \ldots, t_{n-1}) = P(t_n | t_{n-1}) , \qquad (3.61)$$

and the point process is deemed a renewal process.

The opposite case of Eq. (3.60) is when the generation of a spike is considered to be independent of the history of the whole process, so that the firing of a spike is statistically independent of the previous spikes occurrences, such that

$$P(t_n|t_1, t_2, \ldots, t_{n-1}) = P(t_n) .\qquad(3.62)$$

This stochastic process is called a Poisson process. Poisson processes play an important role in the description of spike trains statistics, providing a useful approximation of the stochastic neuronal firing.

As we have seen in Sec. 3.2.2, the firing rate $r(t)$, defined formally by Eq. (3.20), is proportional to the probability of the neuron to fire a single spike around the time t, and, if it is considered that the occurrence of one given spike is independent of the occurrence of other spikes, it can be used to compute the probabilities for all possible action potentials. Specifically, the probability of a neuron firing a spike between the time instants t and $t+\Delta t$ is $P(\text{spike in } [t, t+\Delta t]) = r(t)\Delta t$. Poisson processes can be classified into two main categories: homogeneous Poisson processes, where the firing rate is constant along the spikes' generation, $r(t) = r$, and inhomogeneous Poisson process, where the firing rate is time dependent, $r(t)$.

The inhomogeneous Poisson processes can be further adapted in several different ways to better model the spike train statistics, namely, the inclusion of a model of the absolute refractory period, where the probability of firing is null, and the relative refractory period, where the probability of firing is initially low but increases continually [Berry II and Meister (1998)]. The existence of the refractory period means that the spike events are not independent and that the firing rate by itself is unable to fully describe a spike train correctly.

The majority of neuron models, and retina models in particular, are rate models. This means that their goal is to deliver at the output an estimate of the cell firing rate, but what they convey to the brain is a train of spikes. So, the characterization of the spike train in statistical terms is very important in order to have a model to generate spike trains while departing from an estimation of the firing rate.

3.3.2 Homogeneous Poisson Model of Spike Trains

As we have seen, a spike train is described by the neural response function, $\rho(t)$, with spikes placed at time instants $t_1, t_2, \ldots, t_n$.

First, we divide the spike train with a total time duration T into time bins sufficiently small so that there is at most one spike in each bin. If the width of each of these time bins is Δt, then there are $N = T/\Delta t$ bins; we should recall that the probability of having a spike in a time bin is equal to $r\Delta t$. If we observe a particular spike train with n spikes in the N bins ($n \leq N$), the probability of the occurrence of this very particular sequence is equal to the probability of having n bins with spikes, which is $(r\Delta t)^n$, times the probability of having the other $N - n$ times bins without any spike, which is $(1 - r\Delta t)^{N-n}$. Given the previous reasoning, a particular spike sequence has the probability

$$P(\underbrace{10110\cdots001}_{n \text{ spikes}}) = (r\Delta t)^n (1 - r\Delta t)^{N-n}. \tag{3.63}$$

The argument of $P(\cdot)$ in Eq. (3.63) represents a particular spike train in which we are interested, where a spike is represented by a 1, while an empty bin is represented by 0.

3.3.2.1 Spike Count Distribution

Let us first consider a description of the spike train that disregards the precise time occurrences of each spike. In this description, we are concerned only with the probability of having exactly n spikes within the trial of duration T. We denote this probability by $P_T(n)$, called the *spike count distribution*. If we are not interested in a specific spike train, but instead with the probability of having n spikes within a spike train, independent of the particular time bins in which they appear, then we can have $N!/(n!(N-n)!)$ different spike trains with exactly n spikes distributed among the N bins. Therefore, the probability of getting n spikes in a time period equal to T, for a constant firing rate r, is given by Eq. (3.63) multiplied by the combinatorial factor

$$P_T(n) = \frac{N!}{n!(N-n)!} (r\Delta t)^n (1 - r\Delta t)^{N-n}. \tag{3.64}$$

Equation (3.64) is a discrete probability function, or probability mass function (pmf), corresponding to a binomial distribution. The binomial distribution reduces to the Bernoulli distribution for the particular case of $n = 1$ [Mood et al. (1974); Papoulis and Pillai (2002)], which corresponds to the probability of the occurrence of a single spike during the time period T. Figure 3.25 shows a plot of the probabilities for the number of spikes using the same firing rate $r = 0.3$ spikes/s for different values of the trial duration

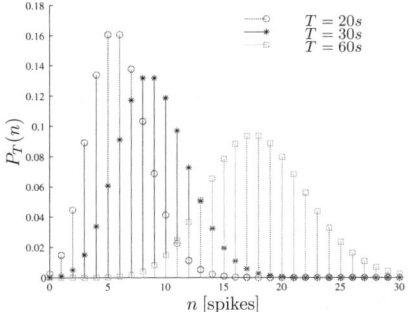
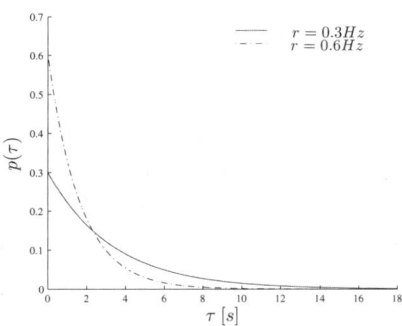

Fig. 3.25 Spike number probability density for a train described by a homogeneous Poisson process with $r = 0.3$ Hz.

Fig. 3.26 Interspike time interval exponential probability density for a spike train described by a homogeneous Poisson process.

T. We can note that, as the value of T increases, the probability of having a larger number of spikes also increases.

In Eq. (3.64), each spike is placed in a given time bin, within the N bins available, which corresponds to a time precision of Δt. However, in continuous terms, the spikes can be placed at any time instant so that we can calculate the limit when the width of the time bins goes to zero. Since $N = T/\Delta t$ as $\Delta t \to 0$, the number of time bins, N, will grow, but $N\Delta t = T$ remains constant, so that Eq. (3.64) becomes:

$$P_T(n) = \lim_{\Delta t \to 0} \frac{N!}{n!(N-n)!} (r\Delta t)^n (1 - r\Delta t)^{N-n}$$
$$= \lim_{N \to \infty} \frac{N(N-1)\cdots(N-n+1)}{n!} \left(r\frac{T}{N}\right)^n \left(1 - r\frac{T}{N}\right)^{N-n}. \tag{3.65}$$

By unfolding the terms in the previous equation, it follows:

$$P_T(n) = \lim_{N \to \infty} \frac{N(N-1)\cdots(N-n+1)}{N^n} \frac{1}{n!} (rT)^n$$
$$\times \left(\left(1 + \frac{-rT}{N}\right)^{\frac{N}{-rT}}\right)^{-rT} \left(1 - r\frac{T}{N}\right)^{-n}. \tag{3.66}$$

Taking the limit and knowing that $\lim_{\epsilon \to 0}(1+\epsilon)^{1/\epsilon} = $ e, we obtain the following equation for the discrete probability density:

$$P_T(n) = \frac{(rT)^n}{n!} e^{-rT}, \tag{3.67}$$

corresponding to a Poisson distribution, where the constant firing rate r appears explicitly.

Two important statistical parameters of a random variable are the mean, or expected value, and the variance. The mean gives the most probable value of the random variable after many observations. For a discrete random variable x, such as the number of spikes in a spike train given by the discrete probability density function (pdf) of Eq. (3.67), with discrete density $P_X(x)$, the mean value, or average, is calculated from

$$\mu_x = E\{x\} = \langle x \rangle = \sum_{k=-\infty}^{+\infty} x_k \, P_X(x_k) \,, \qquad (3.68)$$

where x_k denote the values that the discrete random variable x can take and $P_X(x_k)$ their probability of occurrence. The variance gives a measure of the dispersion of the values of the random variable x around its mean μ_x. It has the definition

$$\begin{aligned}\sigma_x^2 &= \text{VAR}(x) = E\{(x-\mu_x)^2\} \\ &= E\{x^2 - 2\mu_x x + \mu_x^2\} = E\{x^2\} - \mu_x^2 \,,\end{aligned} \qquad (3.69)$$

where $\sigma_x^2 \geq 0$ is a positive quantity. The last equality in Eq. (3.69) results from the fact that the expectation operator $E\{\cdot\}$ defined in Eq. (3.68) is linear. This relation holds both for discrete and continuous random variables and is very useful to calculate the variance. For a discrete random variable, the variance can be computed directly through the expression

$$\sigma_x^2 = \sum_{k=-\infty}^{+\infty} (x_k - \mu_x)^2 \, P_X(x_k) \,. \qquad (3.70)$$

The expected number of spikes, and its variance, within a trial can be calculated using Eq. (3.67) based on Eq. (3.68) and Eq. (3.70), respectively. However, these quantities are more easily calculated with the help of the *moment generating function*. The moment generating function, $M(t)$, for a discrete probability density $P(n)$ is defined as [Papoulis and Pillai (2002)]

$$M(t) = \sum_{n=0}^{+\infty} e^{tn} \, P(n) \,. \qquad (3.71)$$

The moment generating function is useful because the different statistical moments, m_k, can be easily calculated by using the relation:

$$m_k = M^{(k)}(0) \,, \qquad k = 1, 2, \cdots, +\infty, \qquad (3.72)$$

where $M^{(k)}(0)$ is the k-order derivative calculated at the origin:

$$M^{(k)}(0) = \frac{d^k}{dt^k} M(t)|_{t=0} \,. \tag{3.73}$$

The pdf in Eq. (3.67) has the moment generating function:

$$M(t) = \sum_{n=0}^{+\infty} e^{tn} \frac{(rT)^n}{n!} e^{-rT} = e^{-rT} \sum_{n=0}^{\infty} \frac{(rT\,e^t)^n}{n!} \tag{3.74}$$

$$= e^{-rT}\, e^{rT\,e^t} \,.$$

The expected number of spikes, $\langle n \rangle$, in a trial with duration T described by the Poisson pdf in Eq. (3.67), corresponds to the first moment, m_1, of Eq. (3.74):

$$E\{n\} = \langle n \rangle = \sum_{n=0}^{\infty} n P_T(n) = m_1 \tag{3.75}$$

$$= rT \,,$$

which states that the expected number of spikes generated at a constant firing rate is equal to the firing rate r times the considered time interval T. This can be seen in Fig. 3.25. The variance of the spike count in T is

$$\sigma_n^2 = E\{(n - \langle n \rangle)^2\} = \langle n^2 \rangle - \langle n \rangle^2 = m_2 - m_1^2$$
$$= rT \,, \tag{3.76}$$

which is equal to the expected number of spikes. A parameter that characterizes a probability distribution is the Fano factor, defined as

$$F = \frac{\sigma^2}{\mu^2} \,, \tag{3.77}$$

where σ^2 is the distribution variance and μ the mean. The Fano factor characterizes the spike count variability, and, for the case of the Poisson distribution, it takes the value

$$F = \frac{\sigma_n^2}{\langle n \rangle} = 1 \,. \tag{3.78}$$

The fact that the spike count mean and variance are equal is a distinguishing characteristic of a Poisson process. In practice, to analyze if a spike train is adequately described by a Poisson process, one can check if its Fano factor is approximately one [Berry II and Meister (1998)].

3.3.2.2 Interspike Interval Distribution

The probability distribution of the time intervals between adjacent spikes is called the *interspike interval distribution*, and it constitutes another useful characterization of spike patterns.

Given a spike occurring at time t_i, for some value of i, the interspike interval distribution gives the waiting time for the next spike to occur. This probability is equal to the probability that no spikes are generated during a certain interval of time times the probability that a spike is generated in the next time interval.

The probability of not generating a spike in the time interval $[t_i, t_i + \tau]$, of a homogeneous Poisson process, can be obtained by plugging $n = 0$ into Eq. (3.67), which gives

$$P_\tau(n = 0) = e^{-r\tau} . \tag{3.79}$$

The probability of generating a spike in Δt, for a small Δt, is:

$$P_{\Delta t \to 0}(n = 1) = r\Delta t . \tag{3.80}$$

Hence, the probability that a spike fired at instant t_i and that the next spike is generated at t_{i+1}, such that $t_i + \tau \leq t_{i+1} < t_i + \tau + \Delta t$ is:

$$P(\tau \leq t_{i+1} - t_i < \tau + \Delta t) = r\Delta t\, e^{-r\tau} . \tag{3.81}$$

By definition, for a small Δt, the probability density of the interspike intervals is equal to the probability in Eq. (3.81) divided by Δt, which gives

$$p(\tau) = r\, e^{-r\tau} , \tag{3.82}$$

which shows that the interspike-time probability density for a homogeneous Poisson spike train is exponential. As Fig. 3.26 shows, the short interspike intervals are more likely to occur while the long ones have an exponentially decaying probability that is a function of their duration.

An interesting characteristic of the exponential pdf relevant to the description of interspike time interval (ISI) distribution is that it is memoryless, meaning that the time that we have to wait for a new spike is independent of the time that we have already been waiting so far. Let us calculate the probability of a spike occurring in the next period of time $\tau_0 + \tau_1$, given that we already have been waiting for it during τ_0, which means that we want to compute

$$P(t > \tau_0 + \tau_1 | t > \tau_0) . \tag{3.83}$$

By using Bayes' law, this probability is equal to:

$$P(t > \tau_0 + \tau_1 | t > \tau_0) = \frac{P(t > \tau_0 + \tau_1)}{P(t > \tau_0)} . \tag{3.84}$$

Characterization of the Neural Response

To calculate the probabilities in the numerator and denominator of Eq. (3.84), we have to integrate the continuous pdf of Eq. (3.82)

$$P(t > \tau_0 + \tau_1) = \int_{\tau_0+\tau_1}^{+\infty} r\,e^{-r\tau}\,d\tau = e^{-r(\tau_0+\tau_1)}, \qquad (3.85)$$

and

$$P(t > \tau_0) = \int_{\tau_0}^{+\infty} r\,e^{-r\tau}\,d\tau = e^{-r\tau_0}, \qquad (3.86)$$

giving

$$P(t > \tau_0 + \tau_1 | t > \tau_0) = e^{-r\tau_1} = P(t > \tau_1), \qquad (3.87)$$

which means that the fact that no spike was fired during the period τ_0 does not influence the probability of getting one in the next period τ_1. From the interspike interval pdf in Eq. (3.82), we can calculate the mean interspike interval.

For a continuous pdf, $f(x)$, like the ISI density in Eq. (3.82) the mean, or expected value, of the random variable x is defined as

$$E\{x\} = \mu_x = \int_{-\infty}^{+\infty} x f(x)\,dx, \qquad (3.88)$$

and the variance of the random variable x has the definition:

$$\sigma_x^2 = E\{(x - \mu_x)^2\} = \int_{-\infty}^{+\infty} (x - \mu_x)^2 f(x)\,dx. \qquad (3.89)$$

Using the definition of Eq. (3.88), the expected value for the ISI is:

$$E\{\tau\} = \langle \tau \rangle = \int_0^{\infty} \tau r\,e^{-r\tau}\,d\tau = \frac{1}{r}, \qquad (3.90)$$

and, by Eq. (3.89), the variance of the interspike intervals gives

$$\sigma_\tau^2 = \langle \tau^2 \rangle - \langle \tau \rangle^2 = \int_0^{\infty} \tau^2 r\,e^{-r\tau}\,d\tau - \left(\frac{1}{r}\right)^2 = \frac{1}{r^2}. \qquad (3.91)$$

The coefficient of variation, defined as the ratio between the standard deviation, σ, and the mean, μ, is given by

$$C_V = \frac{\sigma}{\mu}. \qquad (3.92)$$

For the interspike interval distribution of a Poisson spike train, it takes the value:

$$C_V = \frac{\sigma_\tau}{\langle \tau \rangle} = 1 . \tag{3.93}$$

Equation (3.93) is a required, but not sufficient, condition to identify a Poisson spike train. For a renewal process, the Fano factor, evaluated over long time intervals, approaches C_V^2 [Dayan and Abbot (2001)].

3.3.3 Inhomogeneous Poisson Model of Spike Trains

As stated in Sec. 3.2.2, the time-dependent firing rate $r(t)$ gives the probability, per unit of time, of a spike occurrence. Equation (3.22) states that the probability of a spike occurring within a time bin of width Δt, at time t, is $r(t)\Delta t$.

Due to the statistical independence of the spike occurrences for a Poisson process, stated by Eq. (3.62), the occurrence probability of a sequence, with spikes placed at time instants $t_1, t_2, \cdots, t_n$, is equal to the probability of finding the spikes in those specific n time bins times the probability of not finding any spike in the remaining $N - n$ time bins ($N = T/\Delta t$).

The probability of finding a spike in a bin with width Δt, centered at the time instant t_i, is $r(t_i)\Delta t$, and the probability of not having a spike at the time bin t_j is $1 - r(t_j)\Delta t$. The probability of the particular spike train occurring with spikes located exactly at times $t_1, t_2, \cdots, t_n$, where $0 \le t_i \le T$, is given by

$$P(t_1, t_2, \ldots, t_n)(\Delta t)^n = \frac{1}{n!} \prod_{j=1; j \ne i}^{N} (1 - r(t_j)\Delta t) \prod_{i=1}^{n} r(t_i)\Delta t$$

$$= \frac{1}{n!} \prod_{j=1}^{N} (1 - r(t_j)\Delta t) \prod_{i=1}^{n} \frac{r(t_i)\Delta t}{1 - r(t_i)\Delta t} , \tag{3.94}$$

where the index j represents all N possible time bins with width Δt. The factor $1/n!$ is introduced because the spikes are indistinguishable and there are $n!$ different ways of assigning the labels. The first product in the second equality of Eq. (3.94) can be simplified by considering that

$$\prod_{j=1}^{N} (1 - r(t_j)\Delta t) = \exp\left(\sum_{j=1}^{N} \ln(1 - r(t_j)\Delta t)\right) , \tag{3.95}$$

and by simplifying the logarithm computation. Since Δt can be made very small, and since, for small x, the Taylor's expansion of the logarithm in the neighborhood of one is $\ln(1+x) = x - 1/2x^2 + 1/3x^3 - \cdots$, then Eq. (3.95) can be approximated to:

$$\prod_{j=1}^{N}(1-r(t_j)\Delta t) =$$

$$= \exp\left(\sum_{j=1}^{N}\left(-r(t_j)\Delta t - \frac{1}{2}(-r(t_j)\Delta t)^2 + \frac{1}{3}(-r(t_j)\Delta t)^3 + \cdots\right)\right)$$

$$= \exp\left(-\sum_{j=1}^{N}r(t_j)\Delta t + \frac{1}{2}\Delta t\sum_{j=1}^{N}r^2(t_j)\Delta t - \frac{1}{3}(\Delta t)^2\sum_{j=1}^{N}r^3(t_j)\Delta t + \cdots\right), \quad (3.96)$$

By applying the property

$$\lim_{\Delta t \to 0}\sum_{j=1}^{T/\Delta t} f(t_j)\Delta t = \int f(t)dt, \quad (3.97)$$

replacing the terms of the exponential of Eq. (3.96), and computing the limit when $\Delta t \to 0$, we arrive at

$$\lim_{\Delta t \to 0}\prod_{j=1}^{N}(1-r(t_j)\Delta t)$$

$$= \lim_{\Delta t \to 0}\exp\left(-\int_0^T r(t)dt + \frac{1}{2}\Delta t\int_0^T r^2(t)dt - \frac{1}{3}(\Delta t)^2\int_0^T r^3(t)dt + \cdots\right)$$

$$= \exp\left(-\int_0^T r(t)dt\right). \quad (3.98)$$

The second product in Eq. (3.94) can also be simplified to:

$$\prod_{i=1}^{n}\left(\frac{r(t_i)\Delta t}{1-r(t_i)\Delta t}\right) = (\Delta t)^n \prod_{i=1}^{n}\left(\frac{r(t_i)}{1-r(t_i)\Delta t}\right) \quad (3.99)$$

and the limit when $\Delta t \to 0$ is:

$$\lim_{\Delta t \to 0}(\Delta t)^n \prod_{i=1}^{n}\left(\frac{r(t_i)}{1-r(t_i)\Delta t}\right) = (\Delta t)^n \prod_{i=1}^{n}r(t_i). \quad (3.100)$$

Replacing the simplifications obtained in Eq. (3.98) and Eq. (3.100), the discrete probability density of Eq. (3.94) becomes

$$P(t_1, t_2, \ldots, t_n) = \frac{1}{n!}\exp\left(-\int_0^T r(t)dt\right)\prod_{i=1}^{n}r(t_i), \quad (3.101)$$

which corresponds to the probability density of a spike train with spikes occurring exactly at the time instants $t_1, t_2, \ldots, t_n$.

The homogeneous Poisson process is a particular case of the inhomogeneous process where the firing rate is held constant during the entire trial duration, $r(t) = r$. The joint probability density for a given number of spikes occurring at specific instants t_i, for a constant firing rate, is

$$P(t_1, t_2, \ldots, t_n) = \frac{1}{n!} e^{-rT} r^n, \qquad (3.102)$$

which is independent of the time occurrence of the spikes. This density can be written as a function of the spike count density, $P_T(n)$, for the homogeneous case given in Eq. (3.67) as:

$$P(t_1, t_2, \ldots, t_n) = P_T(n) \left(\frac{1}{T}\right)^n. \qquad (3.103)$$

3.3.4 Spike-count Statistics

The spike-count distribution characterizes the distribution of the number of spikes from a Poisson model. The joint probability density obtained in expression Eq. (3.101) gives the probability of the occurrence of n spikes at the time instants $t_1, t_2, \cdots, t_n$. To characterize the distribution of the number of spikes along a spike train, Eq. (3.101) must be integrated for all possible time occurrences of the n spikes within T. That is,

$$\begin{aligned} P(n) &= \int_0^T \int_0^T \cdots \int_0^T P(t_1, t_2, \cdots, t_n) dt_1 dt_2 \cdots dt_n \\ &= \int_0^T \int_0^T \cdots \int_0^T \frac{1}{n!} \exp\left(-\int_0^T r(t) dt\right) \prod_{i=1}^n r(t_i) dt_1 dt_2 \cdots dt_n \qquad (3.104) \\ &= \frac{1}{n!} \exp\left(-\int_0^T r(t) dt\right) \int_0^T \int_0^T \cdots \int_0^T \prod_{i=1}^n r(t_i) dt_1 dt_2 \cdots dt_n, \end{aligned}$$

resulting in

$$P(n) = \frac{1}{n!} \exp\left(-\int_0^T r(t) dt\right) \left(\int_0^T r(t) dt\right)^n. \qquad (3.105)$$

It is noted that Eq. (3.67) can be generated from the previous expression considering a constant firing rate equal to r. The average number of spikes can be obtained with a similar procedure used for the homogeneous case, leading to

$$\langle n \rangle = \int_0^T r(t)dt \,, \tag{3.106}$$

which, for the particular case of a constant firing rate r, gives the result presented in Eq. (3.75).

3.3.4.1 *Interspike Interval Distribution*

The interspike interval distribution can be obtained by calculating the probability of not finding a spike during a time interval with length τ, multiplied by the probability of firing a spike at $t = \tau + \Delta t$, with a small Δt. The probability that a spike is not fired during the time τ can be obtained from Eq. (3.105) by making $n = 0$ and assuming that the trial has a duration τ, which results in:

$$P_\tau(0) = \exp\left(-\int_0^\tau r(t)dt\right) . \tag{3.107}$$

As before, the probability of generating a single spike within the small time interval $t \in [\tau, \tau + \Delta t]$ is:

$$P_{\Delta t}(\tau \leq t_i \leq \tau + \Delta t) = r(\tau)\Delta t \,. \tag{3.108}$$

So the interspike interval probability has the following density function:

$$p(\tau) = \exp\left(-\int_0^\tau r(t)dt\right) r(\tau) \,. \tag{3.109}$$

3.4 Spiking Mechanisms

Since many existing retina models produce as their outputs an estimate of the firing rate produced by the RGC, it is important to have a method to generate the spike trains from the firing rate. To generate the spike trains from the firing rate, we can follow the two distinct views already mentioned: we can generate a spike train from the firing rate considering that $r(t)\Delta t$ is the probability of the neuron to fire a spike in the time interval Δt, or we can look to $r(t)\Delta t$ as the number of spikes that we must generate in the time interval Δt. We have seen that the two interpretations given previously are equivalent for a small Δt, but they can lead to two different spike generation mechanisms.

3.4.1 Generation of Poisson Spike Trains

The spike trains are very often considered to be well described by a Poisson process. We will see how to numerically generate a spike train from a variable firing rate $r(t)$ by considering the spike train as a Poisson process. There are two common procedures for generating Poisson spike trains.

A first approach uses the probability of a spike occurring during a short time interval, presented in Eq. (3.22). For a Poisson process, whether it is inhomogeneous or homogeneous, the probability of generating a spike within the small time interval is

$$P\left(n = 1 \text{ in } [t, t + \Delta t]\right) = r(t) \cdot \Delta t . \qquad (3.110)$$

By using this equation, a spike train can be generated by first dividing the time duration of the trial into a sequence of small intervals of width Δt, and also by sampling the firing rate $r(t)$ in intervals of width Δt, composing the sequence $r[n]$. Then, a sequence of random numbers, $x[n]$, is generated with a uniform distribution in the interval $[0, 1]$, and the two sequences are compared. Whenever $r[i]\Delta t \geq x[i]$, a spike is placed at time bin n; otherwise no spike is generated. This procedure is useful when $r[n]\Delta t \ll 1$, and the generated spike trains have a discrete time bin assigned to the spikes. The use of a time bin width of $\Delta t = 1$ ms is usually enough [Dayan and Abbot (2001)].

The second approach used to generate Poisson spike trains with a constant firing rate is to choose the interspike intervals duration randomly from the exponential density for the interspike interval distribution [Berry II and Meister (1998)], given by Eq. (3.82). Each successive spike is placed at a time equal to the previous one plus a random value drawn randomly from the interspike density:

$$t_{i+1} = t_i - \ln(x)/r , \qquad (3.111)$$

where x is a random number generated from a uniform distribution in the interval $[0, 1]$. However, in practice, when the obtained spike train is discretized to serve as the input to the next block of a processing system, this procedure becomes equivalent to the previous one.

To extend this last approach to generate a spike train with a time-varying firing rate, a spike train is generated by considering a constant maximum firing rate $r_{\max}$

$$t_{i+1} = t_i - \ln(x)/r_{\max} , \qquad (3.112)$$

where $r_{\max} > r(t)$, for all t, and x is a random number with uniform distribution in the interval $[0, 1]$. Then, a thinning process is applied to the spike train generated at r_max by keeping or deleting the spike posted at each t_i. The thinning is carried out by generating another random number x with uniform distribution for each i, and if $r(t_i)/r_\text{max} < x$ the spike at time t_i is removed; otherwise it is maintained [Dayan and Abbot (2001)].

The spiking mechanism of a real neuron breaks the assumption of independent firing, as stated by Eq. (3.62), namely during the absolute and relative refractory periods, where the probabilities of firing are null or very low, respectively; these are the principal features of neuronal firing not modeled by a Poisson model. To take into account the refractory effects in the Poisson model, the firing rate can be modulated by a function with zero value just after a spike is fired and during the absolute refractory period, and with an increasing exponential value tending to one during the relative refractory period. The absolute refractory period is variable, and it can be drawn randomly from a Gaussian distribution with a mean equal to the mean absolute refractory period and variance equal to the variance of the absolute refractory period [Dayan and Abbot (2001)].

The Poisson model can be further extended to include refractory periods [Berry II and Meister (1998)] and to model bursting cells that appear predominantly in the visual cortex, where a neuron cell fires a burst of spikes in response to one event [Bair et al. (1994)].

3.4.2 *Integrate-and-Fire Spike Generation*

A straightforward interpretation of the firing rate as the number of spikes fired per second allows for the conversion of $r(t)$ to $\rho(t)$ by generating a spike train with a simple integrate-and-fire method. To code the firing rate $r(t)$ into a set of spikes pulses, as displayed in Fig. 3.27, let us suppose that we start integrating the firing rate, $r(t)$, at the time instant t_i such that

$$v(t) = \int_{t_i}^{t} r(t)dt \ . \tag{3.113}$$

Whenever the activation signal $v(t)$ crosses a predefined threshold ϕ from below, that is when $v(t) \geq \phi$, a spike is fired and the integrator is reset to its rest value

$$\rho(t_i) = \begin{cases} \delta(t - t_i) & v(t_i) \geq 0 \\ 0 & \text{otherwise} \end{cases} . \tag{3.114}$$

Whenever a spike is fired, the feedback loop in Fig. 3.27 is activated and the integrator value is reset to zero ($v(t + \delta t) = 0$), or, alternatively, a function

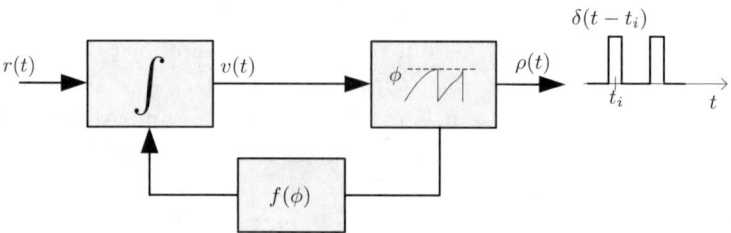

Fig. 3.27 Integrate-and-fire spike generation from firing rate.

of the threshold, $f(\phi)$, with a negative value can be added to it. For the case of a constant firing rate r, we have that $v(t) = r \times (t - t_i)$, and a spike is fired when $v(t)$ reaches the threshold. Considering that the threshold is reached at the time instant t_{i+1}, such that $v(t_{i+1}) = r(t_{i+1} - t_i) = \phi$, we have that the time between two spikes is proportional to the inverse of the firing rate, $t_{i+1} - t_i = \phi/r$. Choosing the right value for the threshold (1 in the previous case), we can generate a sequence of spike trains whose spikes are generated according to the desired firing rate.

The computational implementation of an integrator can be accomplished by taking the relationship between $r(t)$ and $v(t)$ into account, which can be written as

$$r(t) = \frac{dv(t)}{dt}. \tag{3.115}$$

By applying the Euler approximation to the first derivative,

$$\frac{dy(nT_s)}{dt} = \frac{y[(n+1)T_s] - y[nT_s]}{T_s}, \tag{3.116}$$

where T_s is the sampling period used in the discretization process, to Eq. (3.115), we have that

$$r[nT_s] = \frac{v[(n+1)T_s] - v[nT_s]}{T_s}, \tag{3.117}$$

which results in a recursive expression for the computation of $v(t)$ at the time instants $(n-1)T_s, nT_s, (n+1)T_s, \ldots$:

$$v[n+1] = v[n] + r[n]\,T_s \tag{3.118}$$

where the argument dependence on T_s was dropped.

The comparator block in Fig. 3.27 compares $v[n]$ with the threshold, and whenever it crosses ϕ, the output spike sequence is one ($\hat{\rho}[n] = \delta[n] = 1$) and the integrator is reset to zero, or discharged by an amount equal to $f(\phi)$, so that $v[n+1] = 0$, or $v[n+1] = v[n] - f(\phi)$, respectively.

3.5 Conclusions and Further Reading

Formerly, neuron response was measured by counting the number of spikes in a time window applied at the onset of the stimulus presentation; however, currently the experiments are repeated and the nervous cell is stimulated repeatedly with the same stimulus [Rieke et al. (1997)]. The waveform of the evoked potentials has a stereotyped shape for a given class of neuronal cells, and it is common to classify the ganglion cells by the shape of their evoked potentials [Wandell (1995)]. Thus, only their time-stamps are commonly considered in their response analysis, and the spike train is considered to be a point process [Brenner et al. (2002)]. However, this approach is a matter of debate [Eggermont (1998)]. The analysis of such responses relies upon tools from statistical signal processing and probability.

How the spike train represents sensory input, the internal states of the brain and motor control is an open question [Meister and Berry II (1999)]. An interesting illustration used to pose the neural coding/decoding problem is the derided concept of the homunculus [Rieke et al. (1997)].

Owing to the variability of the neurons' responses to the same stimulus, several researchers claim that the important feature of the neural response is just the average firing rate [Berry II and Meister (1998)], which is usually called "rate coding". However, several physiologists, based on recent studies [Uzzell and Chichilnisky (2004)], claim that the response of ganglion cells to the same stimulus is very regular, which indicates that time is relevant in the encoding process [Berry et al. (1997a); de Ruyter van Steveninck and Bialek (1988)]. This "time coding" approach considers that the precise time relations between spikes from different neurons are meaningful. On the other hand, some studies have shown that 90% of the information is coded by a ganglion cell alone, and only 10% of the information is encoded by the population response [Nirenberg et al. (2001)].

For a more extensive presentation and analysis of the tools used in computational neuroscience see [Dayan and Abbot (2001)]. For a detailed treatment and further extensions of the probabilistic analysis of the neural response, we strongly recommend reading [Rieke et al. (1997)].

Exercises

3.1. *A continuous random variable has an exponential distribution if its pdf has the form*

$$p(x) = \lambda e^{-\lambda x}, \qquad (3.119)$$

where λ is a parameter of the distribution (see Fig. 3.26 for $\lambda = r$). To generate a sequence of samples with an exponential distribution we can use the inverse transform method. The distribution function for the exponential density is given by:

$$\begin{aligned} F(x) = P(X \leq x) &= \int_0^x \lambda e^{-\lambda \tau} \, d\tau \\ &= 1 - e^{-\lambda x} \quad 0 < x < \infty \, . \end{aligned} \qquad (3.120)$$

By making

$$u = F(x) = 1 - e^{-\lambda x}, \qquad (3.121)$$

and solving for x we get

$$x = -\frac{1}{\lambda} \ln(1 - u) \, . \qquad (3.122)$$

By noting the fact that $1 - u$ is uniformly distributed over the interval $(0, 1)$ we can generate exponential random variables with parameter λ using the transformation

$$X = -\frac{1}{\lambda} \ln(Z) \, , \qquad (3.123)$$

where Z is a random number uniformly distributed over the interval $(0, 1)$. The generating process is described in Algorithm 3.1.

Algorithm 3.1 Generation of exponential random variable sequence

---
1: $N \leftarrow$ number of samples
2: $\lambda \leftarrow$ parameter of the exponential distribution
3: $\boldsymbol{X} \leftarrow N$ dimensional column vector of generated samples
4: $P_0 = e^{-\lambda}$
5: **for** $j = 1$ to N **do**
6:   Generate a uniform random number Z in the interval $(0,1)$
7:   $\boldsymbol{X}_{j,1} = -\ln(Z)/\lambda$
8: **end for**
9: **return** $\boldsymbol{X}$

3.1.1 *Generate a sequence of random numbers (e.g., $N = 2000$ samples) with an exponential distribution, where the parameter λ can be chosen arbitrarily. Plot the histogram of the results along with the theoretical probability density function of Eq. (3.119).*

3.1.2 *Calculate the sample mean, variance and the coefficient of variation and compare with the theoretical results.*
The sample mean is given by:

$$\bar{x} = \frac{1}{N} \sum_{i=1}^{N} x_i \,, \qquad (3.124)$$

where N is the number of samples and x_i the value of sample i. The sample variance is given by expression:

$$\mathrm{VAR}(x) = \frac{1}{N-1} \sum_{i=1}^{N} (x_i - \bar{x})^2 \,, \qquad N > 1 \,. \qquad (3.125)$$

3.2. *A discrete random variable x has a Poisson distribution if its density function is given by*

$$P_X(x) = \frac{\mathrm{e}^{-\lambda} \lambda^x}{x!} \qquad x = 0, 1, 2, \ldots \,. \qquad (3.126)$$

Using the following recursive relationship between successive Poisson probabilities:

$$P_X(x = i+1) = \frac{\lambda}{i+1} P_X(x = i) \,, i \geq 0 \quad \text{where} \quad P_X(x = 0) = \mathrm{e}^{-\lambda} \,. \qquad (3.127)$$

we can generate values for a random variable, X, with a Poisson distribution by the inverse transform method [Ross (2006)] using the Algorithm 3.2.

3.2.1 *Develop a program based on Algorithm 3.2 to generate a sequence of random numbers (e.g., N=2000 samples) with a Poisson distribution where the parameter λ can be chosen arbitrarily.*

3.2.2 *Predict the values for the mean and for the variance of the sequence using the definitions in Eq. (3.68) and (3.70).*

3.2.3 *Compute the sample mean and the sample variance of the generated sequence (see Exercise 3.1). Observe the differences of the mean and variance for different values of the sequence size. Compare the calculated results with the ones obtained in the previous question.*

3.2.4 *Calculate the Fano factor of the generated sequence. What can you conclude?*

Algorithm 3.2 Generation of Poisson random variable sequence

1: $N \leftarrow$ number of samples
2: $\lambda \leftarrow$ Poisson distribution parameter
3: $\boldsymbol{X} \leftarrow N$ dimensional column vector of generated samples
4: $P_0 = e^{-\lambda}$
5: $j = 1$
6: **while** $j \leq N$ **do**
7:     Generate a uniform random number Z in the interval (0,1)
8:     $i = 0$
9:     $F_0 = P_0$
10:    flag = TRUE
11:    **while** flag = TRUE **do**
12:       **if** $Z \leq F_i$ **then**
13:          $\boldsymbol{X}_{j,1} = i$
14:          flag = FALSE
15:          $j = j + 1$
16:       **else**
17:          $P_{i+1} = \lambda P_i / (i+1)$
18:          $i = i + 1$
19:          $F_{i+1} = F_i + p_{i+1}$
20:       **end if**
21:    **end while**
22: **end while**
23: **return** $\boldsymbol{X}$

3.3. *The interspike interval distribution follows a continuous exponential density, as stated by Eq. (3.82), for a homogeneous Poisson spike train.*

   3.3.1 *Develop a Matlab program to generate a set of spike times with 20 trials, with a time duration for each trial of 100 s, for different values of the constant firing rate r.*
   Suggestion: Use the procedure described in Exercise 3.1, and by making note that the time between the spikes placed at t_i and t_{i+1} is $t_{i-1} - t_i = \tau$, so that $t_{i+1} = t_i - \ln(Z)/r$ (where Z is a uniformly distributed random variable in the interval $(0,1)$).
   3.3.2 *Using the output of the program developed in the previous exercise obtain the respective spike trains from the spike time sequences, ob-*

tained with a constant firing equal to $r = 8$ spikes/s, for the sampling periods $T_s = 1$ ms, $T_s = 10$ ms and $T_s = 100$ ms. (It is better to write a Matlab program to do this automatically!)

3.3.3 Repeat the previous exercises but using a variable firing rate. The firing rate, $r(t)$ can be generated from a uniform distribution with the range $(0, 1)$ (you can use other probability distributions but keep the range of values between 0 and 1 since $r(t)$ is a probability!).

3.4. Given a set of spike trains, develop a Matlab script to calculate the spike-count rate, the firing rate, and the average firing rate, using:

3.4.1 A rectangular window with width chosen by the user.

3.4.2 A Gaussian window with a variable width, given by the parameter σ, chosen by the user.

3.4.3 A variable rectangular window with a fixed number of spikes.

Suggestion: If you don't have access to real spike trains you can always generate a set of trials following the procedure described in Exercise 3.3.

3.5. With a set of spike trains, calculate the firing rate using several different widths for the time bin Δt. For example, use $\Delta t = T_s$, $\Delta t = 50T_s$, and $\Delta t = 100T_s$, where T_s is the sampling period used to discretize the spike trains. Compare the values obtained when you integrate (sum in discrete-time) the firing rate and multiply the result by the bin width. (If you don't real spike trains generate a set of trials following the procedure described in Exercise 3.3.)

3.6. Develop the full mathematical calculations to show the equalities displayed in Eq. (3.76) and Eq. (3.75) corresponding to the mean and variance of the spike count discrete density given by Eq. (3.67).

3.7. Following a procedure similar to the homogeneous case, show that the average number of spikes for a spike train described by a nonhomogeneous Poisson process is given by Eq. (3.106).

3.8. Write a Matlab script that implements a Poisson spike generator.

3.8.1 Generate a set of spike trains with constant firing rate chosen arbitrarily by the user.

3.8.2 Calculate the mean firing rate from the generated trains and compare with the original for different time lengths of the train.

3.8.3 Calculate the Fano factor for the Poisson spike trains generated.

3.9. Using the Poisson spike trains generated in Exercise 3.8, write a Matlab program to obtain the ISI distribution. Compare the results obtained from an exponential distribution with the appropriate firing rate.

3.10. Implement a Poisson spike generator that generates a spike train from a discrete sequence of firing rates.

3.11. Implement an integrate-and-fire spike generator to generate spike trains from a discrete sequence of firing rates.

3.12. Compare the spike trains, in terms of the mean number of spikes and its variance, for the two methods of spike generation implemented in Exercise 3.10 and Exercise 3.11. Compute the Fano factor with the results from the two spike generation methods.

Chapter 4

Retina Models

4.1 Introduction

Modeling a neural processing system is a major goal in neuroscience. It allows additional knowledge about how the brain and the central nervous system work to be obtained. Moreover, it enables the development of electronic devices to overcome impairments, such as a bioelectronic vision system for the blind, or to extend our natural senses.

Our main goal in the previous chapters was to uncover the neural code so that it would be possible to predict the sequence of spikes fired by a neuron in the visual system when excited with a light stimulus. In this chapter, we outline several representative retina models of different model classes, both in continuous and discrete-time forms, describing the necessary steps for a computational implementation targeting a bioelectronic vision device.

4.2 Classification of Retina Models

The first major division where neural models can be classified is between rate-code models and time-code models. In the case of the retina, the output of a rate-code model is just the mean firing rate of the RGC given the light stimulus, while for a time-code model its output corresponds to the spike train relative to the encoding of the gathered image.

In terms of its internal structure, and also from a macroscopic point of view, retina models can be classified as structural or functional. In structural models, the cell functional structures are modeled normally with a set of differential equations. More recently, there is a tendency to model each retina cell layer individually to obtain a complete retina model by cascading these individual structures. On the other hand, functional retina models treat the retina as a black box and just try to resemble its functioning. Despite their simplicity, these models cannot be ignored, as they can mimic the retinal response.

Within the functional model class, we can distinguish several families. The most evident distinction is between deterministic and stochastic models. Deterministic models, normally of the rate-code type, always produce the same output response for a given input stimulus. On the other side, stochastic models try to simulate the intrinsic variability of the retinal response by introducing noise sources into the model's internal structure, and may directly produce the spike trains. Amid these, there is the relevant class of white noise models, derived by applying white noise analysis techniques to the retina's responses. In the following sections, we will provide examples of models from each of these classes that have been thoroughly investigated.

Generally, neural models rely on the assumption that neurons generate spikes independently of each other. Another assumption is that the neuron's firing probability varies in time and is exclusively a function of the input stimulus. These assumptions are not entirely true, as the neurons compose an intricate network with multiple feedforward and feedback paths, even in the primary neuronal layers of the retina. Moreover, a neuron cannot fire two spikes with an arbitrarily small interspike interval, because their internal and external ion levels must be recovered in order to be able to fire again, which leads to the concepts of absolute and partial refractory periods [Berry II and Meister (1998)]. Several models appear in the literature that take these phenomena into account by modifying their structure.

Every model of the retina tries to map the light pattern incident on the photoreceptors' layer to the spike train that comes out from the ganglion cells' layer. This light pattern, focused on the retina by the eye's optical system, is described as a function of three variables, corresponding to: i) the two spacial dimensions vector, $\mathbf{r} = [x\,y]^T$, ii) as a function of time, t, that takes into account the changing of environment components, by itself or due to the eyes' movements, and iii) and as a function of light wavelength, λ. The third spatial dimension, depth, is perceived by the brain through the angular difference between the two two-dimensional images gathered by each eye. Hence, the visual stimulus signal can be described mathematically by the function $s(\mathbf{r}, t, \lambda)$. Normally, retina models treat the stimulus only as a function of space and time, so that its dependence on wavelength is deferred, or treated separately for each wavelength. The light intensity patterns can also be continuous or discrete, in space and/or in time.

Although the surrounding visual stimuli are spatiotemporal, the retina response is frequently only modeled in time, and even in the spatiotemporal models it is common to consider spatial and temporal processing separable, and therefore treated independently [Wandell (1995)].

4.3 Structural Models

A family of neuron models with wide application in the modeling of different neural processing centers, including retina nervous cells, are the integrate-and-fire (I&F) models [Reich et al. (1997); Gestri et al. (1980)]. These models are based on the characteristics of the neuronal membrane, whose properties can be modeled as electric components, and are a simplification of the Hodgkin-Huxley nonlinear set of differential equations [Gerstner and Kistler (2002)].

4.3.1 *The Integrate and Fire Model*

The simplest structural model of a neuron assumes that it integrates the input stimulus and fires a spike whenever the membrane voltage, $V_m(t)$, surpasses a threshold voltage, V_θ, from below. The general integrate-and-fire model is depicted in Fig. 4.1.

The simple integrate-and-fire model considers that the neuronal membrane behaves like a capacitor that is charged with the input stimulus current until it reaches a limit value, V_θ, and discharges to a reset value, V_r. This behavior is equivalent to considering a very high membrane resistance ($R_m \to \infty$) in Fig. 4.1, which is equivalent to an open circuit, and by regarding the switch internal resistance as very small ($R_{int} \to 0$), which is equivalent to a short circuit. The relation between the voltage, $V_m(t)$, across the neuron's membrane capacity, C_m, and the input stimulus current, $I_s(t)$, is given by

$$C_m \frac{dV_m(t)}{dt} = I_s(t) . \tag{4.1}$$

Equation (4.1) states that the input current $I_s(t)$ is integrated to yield the membrane voltage $V_m(t)$. Whenever $V_m(t)$ crosses the threshold potential V_θ a spike is fired and the switch is closed, resetting the membrane potential to the reset value, V_r.

This integrate-and-fire model is characterized by possessing an infinite memory so that even a very small current applied for a sufficiently long

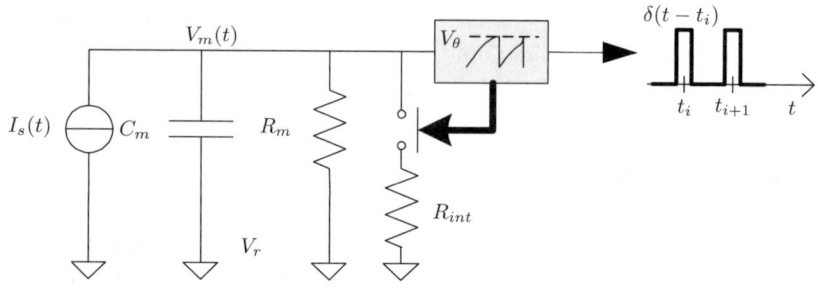

Fig. 4.1 The integrate-and-fire (I&F) model.

period will trigger a spike, and a constant firing rate is produced for any nonzero constant input current.

The membrane potential, $V_m(t)$, can be obtained by solving the differential equation in Eq. (4.1) by integration. Integrating Eq. (4.1) between the time instants t_i and t, we obtain

$$V_m(t) = V_m(t_i) + \frac{1}{C_m} \int_{t_i}^{t} I_s(t)dt, \qquad t_i \leq t < t_{i+1}. \qquad (4.2)$$

If we state that the neuron fired a spike at the time instant t_i, its membrane is at the reset potential, $V_m(t_i) = V_r$. From Eq. (4.2), it follows that

$$V_m(t) = V_r + \frac{1}{C_m} \int_{t_i}^{t} I_s(t)dt, \qquad t_i \leq t < t_{i+1}. \qquad (4.3)$$

Equation (4.3) represents the membrane potential excursion for $t_i \leq t < t_{i+1}$, where t_{i+1} is the time instant of the next generated spike after the spike in t_i.

It is illustrative to see the effect of a constant input current. For a constant stimulus input current, $I_s(t) = I_s$, from Eq. (4.3), the membrane potential is given by

$$V_m(t) = V_r + \frac{1}{C_m} I_s(t - t_i), \qquad t_i \leq t \leq t_{i+1}. \qquad (4.4)$$

This expression is valid between the time instant when the neuron fired a spike, $t = t_i$, until the firing of the next spike, $t = t_{i+1}$, when the membrane is set to the reset potential and its voltage follows Eq. (4.4) once again. Since a neuron fires a spike whenever $V_m(t)$ surpasses V_θ from below, the

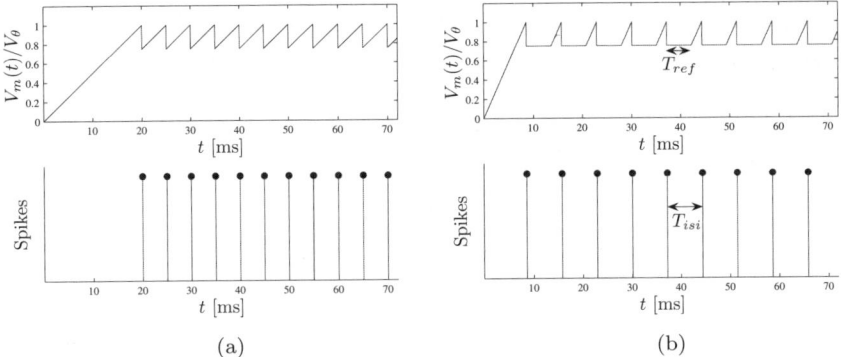

Fig. 4.2 Integrate-and-fire model's responses for a constant input stimulus current. *Top*: Normalized membrane potential. *Bottom* Generated spike sequence. (a) without refractory period (b) and with a refractory period. ($I_s = 1.5$ mA, $C_m = 1$ nF, and $T_{ref} = 5$ ms; ($V_r = 45$ mV, $V_\theta = 60$ mV)).

time between two consecutive spikes, T_{isi}, can be obtained from Eq. (4.4) by knowing that $V_m(t_{i+1}) = V_\theta$, which gives

$$T_{isi} = t_{i+1} - t_i = C_m \frac{V_\theta - V_r}{I_s}. \tag{4.5}$$

From the time between spikes, the firing rate, $r = 1/T_{isi}$, becomes

$$r = \frac{1}{C_m} \frac{I_s}{V_\theta - V_r}, \tag{4.6}$$

which is constant because the time between spikes is also constant.

Figure 4.2(a) displays the time evolution of the membrane voltage for a constant input current, and displays the corresponding spike sequence generated by the model. Note that by adjusting the threshold potential to $V_\theta - V_r$, the reset potential in Fig. 4.1 can be set to zero ($V_r = 0$).

Equation (4.6) states that the neuron can fire at an arbitrarily high frequency for a high input stimulus current, I_s. However, in real neurons this situation cannot occur due to the dynamics of the ion channels across the membrane, which introduces a refractory period after every firing. A neuron that has fired a spike is unable to produce another for a period of time T_{ref}, regardless of the strength of the stimulus. After this dead time, in the absence of refractoriness, its firing ability returns immediately to its prior ready state. The integrate-and-fire model can be extended to take into account the refractory period by stating that after a fire, the neuron is inactive during a given time T_{ref}, and only after this period will it be able to fire again. Referring to Fig. 4.1, the refractory period can be modeled by stating that the switch is closed during a time equal to T_{ref}, and only after this period it is open again and the capacitor can start charging to fire a new spike.

For the case of a constant input stimulus current, taking into account the refractory period T_{ref}, the time between two consecutive spikes is

$$T_{isi} = T_{ref} + \frac{C_m V_\theta}{I_s}, \qquad (4.7)$$

which corresponds to the firing rate

$$r = \frac{I_s}{C_m V_\theta + T_{ref} I_s}. \qquad (4.8)$$

From Eq. (4.8), we observe that for large stimuli currents (meaning that the neuron is being heavily excited) the maximum firing rate is

$$r_{max} = \lim_{I_s \to \infty} \frac{I_s}{C_m V_\theta + T_{ref} I_s} = \frac{1}{T_{ref}}, \qquad (4.9)$$

which is limited by T_{ref}. Figure 4.2(b) displays the time evolution of the membrane voltage for the integrate-and-fire model considering a refractory period after the firing of each spike. At the beginning ($t = 0$ s), the neuron is at its resting potential ($V_{rest} = 0$ V), and it integrates the input stimulus current until it reaches the threshold V_θ, where a spike is fired and the membrane potential is set to its reset value, V_r. Figure 4.4 displays the evolution of the firing rate as a function of the stimulus current for the integrate-and-fire model with, and without, refractory period.

4.3.2 The Leaky Integrate-and-Fire Model

The leaky integrate-and-fire (LI&F) model includes a leak term that models the fact that the neuron's memory is not infinite, as it was considered in the previous integrate-and-fire model, so that past input stimuli events are discarded as time goes by.

The leak term models the current drain from the cell's membrane, and is included in the differential equation Eq. (4.1) like:

$$C_m \frac{dV_m(t)}{dt} = I_s(t) - \frac{V_m(t)}{R_m}. \qquad (4.10)$$

This leak current, $I_{leak}(t)$, is proportional to the quotient of the membrane voltage by the membrane resistance, R_m, following Ohm's law. The leaky integrate-and-fire model is represented by the electric circuit of Fig. 4.1, where the resistance R_m is placed in parallel with a capacitor C_m that resembles the membrane capacity.

By introducing the time constant $\tau_m = R_m C_m$, the differential equation for the leaky integrate-and-fire model takes the form:

$$\tau_m \frac{dV_m(t)}{dt} + V_m(t) = R_m I_s(t), \qquad (4.11)$$

where time constant τ_m characterizes the membrane dynamics. If the input current is unable to recover the leak current, the membrane is driven to its potential rest value, which for this case is zero ($V_{rest} = 0V$), and consequently, the neuron will never fire a spike.

For the case of a constant input stimulus current, I_s must be strictly bigger than V_θ/R_m, so that it can compensate for the maximum drain current, $I_\theta = V_\theta/R_m$, in order for the neuron to fire. We examine the model behavior for a constant input current $I_s(t) = I_s$, such that Eq. (4.11) becomes

$$C_m \frac{dV_m(t)}{dt} + \frac{V_m(t)}{R_m} = I_s. \qquad (4.12)$$

The solution of this differential equation is composed of the sum of the homogeneous, or free, solution, which is

$$V_{m_h}(t) = K e^{-\frac{t}{\tau_m}}, \qquad (4.13)$$

with the particular, or forced, solution for a constant current input that is given by

$$V_{m_p}(t) = R_m I_s, \qquad (4.14)$$

so that the overall solution is $V_m(t) = V_{m_h}(t) + V_{m_p}(t)$. The constant of integration, K, in Eq. (4.13) is obtained by stating that at the initial time instant t_i, the neuron fires so that its membrane potential is at the reset value $V_m(t_i) = V_r$. The overall solution of Eq. (4.12) is:

$$V_m(t) = V_r e^{-\frac{t-t_i}{\tau_m}} + R_m I_s \left(1 - e^{-\frac{t-t_i}{\tau_m}}\right). \qquad (4.15)$$

The obtained solution can be checked by substituting it back into Eq. (4.12) and verifying the equality.

Knowing that a spike occurs whenever $V_m(t) = V_\theta$, we can calculate from Eq. (4.15) the time of the next spike occurrence. Making $V_m(t_{i+1}) = V_\theta$ in Eq. (4.15) we obtain the interspike interval

$$t_{i+1} - t_i = \tau_m \ln\left(\frac{V_r - R_m I_s}{V_\theta - R_m I_s}\right). \qquad (4.16)$$

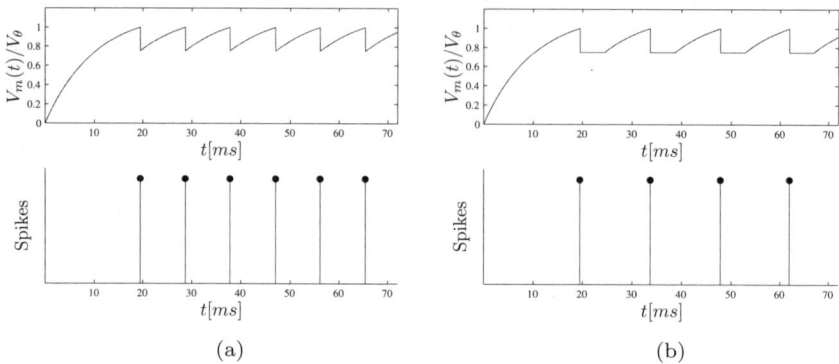

Fig. 4.3 Leaky integrate-and-fire model's response for a constant input stimulus current. *Top*: Normalized membrane potential. *Bottom* Generated spike sequence. (a) without refractory period (b) and with a refractory period. ($I_s = 7$ nA, $C_m = 1$ nF, $R_m = 10$ MΩ, and $T_{ref} = 5$ ms; ($V_r = 45$ mV, $V_\theta = 60$ mV)).

By taking into account a refractory period of T_{ref}, the overall ISI is $T_{isi} = T_{ref} + (t_{i+1} - t_i)$, and the firing rate for the leaky integrate-and-fire model with a constant input current becomes

$$r = \begin{cases} 0 & I_s \leq I_\theta \\ \left[T_{ref} + \tau_m \ln\left(\frac{V_r - R_m I_s}{V_\theta - R_m I_s}\right)\right]^{-1} & I_s > I_\theta \end{cases}. \quad (4.17)$$

Figure 4.3(b) displays the model's membrane potential for the leaky integrate-and-fire model stimulated by a constant input current with a refractory period, and Fig. 4.4 displays the evolution of the firing rate as a function of the stimulus current for the leaky integrate-and-fire model with, and without, refractory period.

To find the form of the general solution to Eq. (4.10) for any type of input stimuli, we can recall the appropriate integrating factor from differential equation theory. By multiplying both sides of Eq. (4.10) by e^{t/τ_m} we get

$$e^{\frac{t}{\tau_m}} \left(\frac{dV_m(t)}{dt} + \frac{V_m(t)}{\tau_m}\right) = \frac{1}{C_m} e^{\frac{t}{\tau_m}} I_s(t). \quad (4.18)$$

Integrating the left hand side of Eq. (4.18):

$$\int_{t_i}^{t} e^{\frac{x}{\tau_m}} \left(\frac{dV_m(x)}{dx} + \frac{1}{\tau_m} V_m(x)\right) dx = e^{\frac{x}{\tau_m}} V_m(x)\Big|_{t_i}^{t}$$

$$= e^{\frac{t}{\tau_m}} V_m(t) - e^{\frac{t_i}{\tau_m}} V_m(t_i). \quad (4.19)$$

Stating once again that the neuron has fired a spike at time instant t_i and the neuron membrane's potential is $V_m(t_i) = V_r$, and equating with the integration of the right hand side of Eq. (4.18), we get

$$V_m(t) = e^{-\frac{t-t_i}{\tau_m}} V_r + \frac{1}{C_m} \int_{t_i}^{t} e^{-\frac{t-x}{\tau_m}} I_s(x) dx, \quad t \geq t_i . \tag{4.20}$$

Equation (4.20) can be rewritten by using the Heaviside unit step function $u(t-x)$ into the integrand, which is reversed and shifted so that $u(t-x) = 0$ for $x > t$, and taking into account that after the firing of a spike at t_i the input stimulus current before instant t_i does not affect the neuron for $t > t_i$. Thus, $I_s(t)$ is multiplied by $u(t-t_i)$

$$V_m(t) = e^{-\frac{t-t_i}{\tau_m}} V_r + \frac{1}{C_m} \int_{-\infty}^{\infty} u(t-x) e^{-\frac{t-x}{\tau_m}} I_s(x) u(x-t_i) dx, \quad t \geq t_i . \tag{4.21}$$

In Eq. (4.21) the integral corresponds to a convolution operation, so that the final expression for the general solution of Eq. (4.10) is

$$V_m(t) = e^{-\frac{t-t_i}{\tau_m}} V_r + \frac{1}{C_m} [I_s(t) u(t-t_i)] * \left[u(t) e^{\frac{-t}{\tau_m}} \right], \quad t \geq t_i . \tag{4.22}$$

The first term on the right side of Eq. (4.22) can also be expanded, so that the expression for the membrane potential after a firing at t_i is

$$V_m(t) = V_r + \left[\frac{1}{C_m} I_s(t) u(t-t_i) - \frac{V_r}{\tau_m} u(t-t_i) \right] * \left[u(t) e^{\frac{-t}{\tau_m}} \right], \quad t \geq t_i . \tag{4.23}$$

After the firing of a spike, the membrane potential is given by the convolution of the input stimulus current with a low-pass filter plus an exponentially decaying term, which in many situations and for modeling purposes has a null contribution by making $V_r = 0$. Note that Eq. (4.23) is valid from the firing of a spike (time instant t_i) until the next spike is fired (time instant t_{i+1}). Equation (4.23) can be used repeatedly to obtain the next firing instant by considering the previous time instant t_{i+1} as the new t_i.

Computational Implementation For the computational implementation of the leaky integrate-and-fire model we need to find the discrete counterpart of the differential in Eq. (4.11). In the discrete implementation we want to find the membrane voltage, $V_m(t)$, for the time instants $t = nT_s$, where $n = 0, 1, 2, \ldots$, and T_s is the sampling period. That is, we want to find $V_m[0], V_m[T_s], V_m[2T_s], \ldots, V_m[nT_s], V_m[(n+1)T_s], \ldots$. The simplest

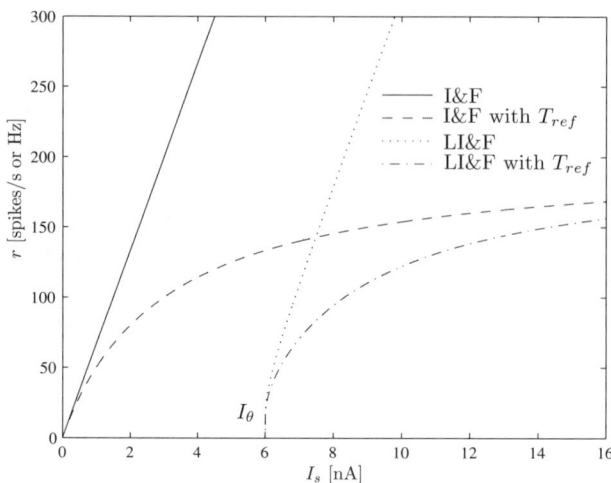

Fig. 4.4 Firing rate versus stimulus current for the integrate-and-fire models. ($I_s = 7$ nA, $C_m = 1$ nF, $R_m = 10$ MΩ, and $T_{ref} = 5$ ms).

way to obtain the discrete form of Eq. (4.11) is to approximate the continuous derivative by the forward discrete approximation, also known as the Euler approximation:

$$\frac{dy(nT_s)}{dt} = \frac{y[(n+1)T_s] - y[nT_s]}{T_s} . \qquad (4.24)$$

This approximation improves as the sampling period T_s gets smaller; in the limiting condition of $T_s \to 0$, we obtain the original function. Applying this approximation, and dropping the dependency on the sampling period in the arguments, to the differential equation in Eq. (4.11), it becomes

$$\tau_m \frac{V_m[(n+1)] - V_m[n]}{T_s} + V_m[n] = R_m I_s[n] , \qquad (4.25)$$

in discrete-time. By applying basic mathematical manipulations Eq. (4.25) becomes:

$$V_m[n+1] = \left(1 - \frac{T_s}{\tau_m}\right) V_m[n] + \frac{T_s}{C_m} I_s[n] . \qquad (4.26)$$

This equation can be directly computed. The membrane potential should be initialized with the rest potential, which is assumed to be zero, $V_m[0] = 0$ V, and whenever it crosses the threshold potential, when $V_m[n] \geq V_\theta$, a spike is fired and its value is set to the reset potential, $V_m[n+1] = V_r$.

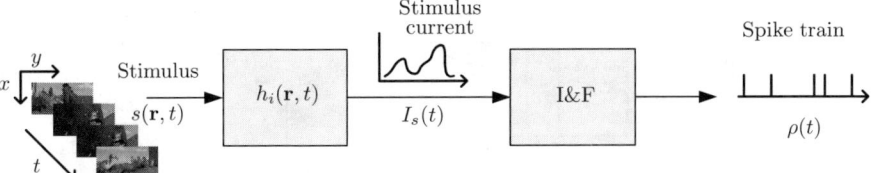

Fig. 4.5 Block diagram of the integrate-and-fire model of the retina.

Finally, we need to reveal where the input stimulus current $I_s(t)$ comes from. The stimulus current comes from the neuron's synaptic connections with the neighboring neurons. In the first layer of the retina, we can consider that the stimulus current comes from the photoreceptors' perception of the input light pattern and can be obtained by the convolution of the visual stimuli with a filter or with a bank of filters. If we are just interested in modeling the RGC, then the shape of the filter resembles its temporal receptive field; it can be, for example, the STA presented in Section 3.2.3. It is common to use a set of basis functions (e.g., Laguerre basis functions, or distorted sine basis functions as represented in Fig. 4.9) to model this input filter. This can be viewed as the receptive field of the neuron that selects the relevant features from the stimulus to fire a spike. Figure 4.5 shows a block diagram of the integrate-and-fire model of a neuron with the input stimulus current generator block.

For a general input stimulus with spatial and temporal variations, the input stimulus current is obtained through the expression:

$$I_s(t) = \sum_{i=0}^{N} h_i(x,y,t) * s(x,y,t) , \qquad (4.27)$$

where $*$ represents the convolution operator that is accomplished in space and in time. The sum index N is the number of basis functions composing the impulse response of the input filter.

4.4 Functional Models

The following sections are devoted to the description of several functional models of the retina. These models are classified according to the type of processing they use to estimate the neural code of the RGCs. The functional models presented next are tuned and assessed in Chap. 5.

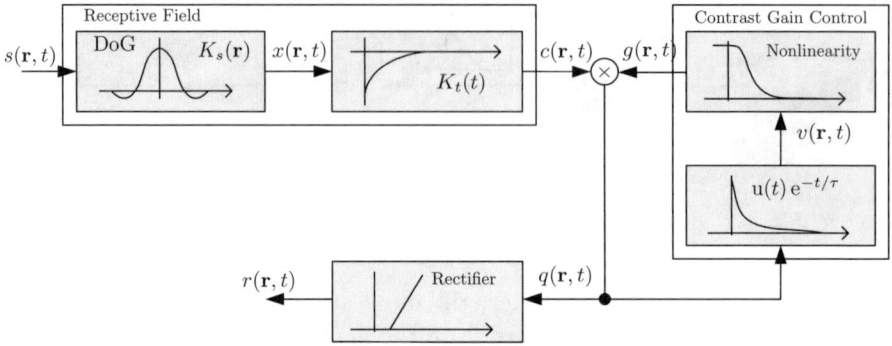

Fig. 4.6 Block diagram of a deterministic model.

4.4.1 *Deterministic Models*

The deterministic models were given their name because they do not model the variations in the retina response to a given stimulus. Using the same stimulus as input they will always produce the same output response.

The deterministic retina model described and analyzed next was reported in [Wilke *et al.* (2001); Thiel *et al.* (2003)] and outputs the firing rate. It was developed to model the temporal and spatial response of ON and of ON-OFF-type ganglion cells of a turtle retina to the movement of a white bar crossing its visual field. It was observed that the temporal behavior of these two kinds of neurons are similar [Wilke *et al.* (2001)]. The difference is mainly noted in their spatial response, where the ON-type ganglion cells respond only to the onset of the bar, whereas the ON-OFF-type cells respond to the onset and to the offset of the light bar.

Model Description The block diagram of this model is depicted in Fig. 4.6. The signal $s(\mathbf{r}, t)$ represents the light stimulus pattern that hits the retina as a function of time t and space $\mathbf{r}$, where the vector $\mathbf{r}$ represents the stimulus spatial dependence, $\mathbf{r} = [x\,y]^T$. The first block of the model resembles the RF of the RGC. The spatiotemporal stimulus pattern is convolved with a kernel, $K(\mathbf{r}, t)$, resulting in the activation signal of the ganglion cell, $c(\mathbf{r}, t)$. This operation is described by the convolution

$$c(\mathbf{r}, t) = K(\mathbf{r}, t) * s(\mathbf{r}, t) \,. \tag{4.28}$$

In a real retina, the spatiotemporal filter $K(\mathbf{r}, t)$ may have a time delay: $K(\mathbf{r}, t) = \tilde{K}(\mathbf{r}, t+\delta t)$. For typical RFs of RGCs this kernel can be factorized, within a good approximation, in a time and a space kernel [Wandell (1995)],

such that
$$K(\mathbf{r}, t) = K_s(\mathbf{r})K_t(t) . \qquad (4.29)$$

To model the spatial RF of a RGC, a difference of Gaussians (DoG) [Wandell (1995)] is frequently used [Rodieck (1965)]. A DoG has the mathematical description:

$$K_s(\mathbf{r}) = \frac{A_C}{2\pi\sigma_C^2} \exp\left(-\frac{\mathbf{r}^2}{2\sigma_C^2}\right) - \frac{A_S}{2\pi\sigma_S^2} \exp\left(-\frac{\mathbf{r}^2}{2\sigma_S^2}\right) ; \qquad (4.30)$$

where the parameters A_C and A_S give the weight of the center of the RF relative to its surroundings, and the parameters σ_C^2 and σ_S^2 ($\sigma_C^2 < \sigma_S^2$), control the diameter of the center and of the outer Gaussian functions, respectively. The spatial behavior of the RF of the ON and OFF ganglion cells can be modeled by carefully choosing these weights. The effect of a DoG is a band-pass spatial filtering.

The temporal kernel corresponds to a high-pass filter, typically expressed by the following equation

$$K_t(t) = \delta(t) - \alpha\, \mathrm{u}(t)\, e^{-\alpha t} , \qquad (4.31)$$

where $1/\alpha$ is the decay rate of the filter response and $\mathrm{u}(t)$ is the continuous Heaviside unit step function.

The feedback loop of the model consists of a CGC block [Berry II et al. (1999)], inserted to capture the characteristics of the retinal response. The signal $c(\mathbf{r}, t)$ is multiplied by $g(\mathbf{r}, t)$, giving

$$q(\mathbf{r}, t) = c(\mathbf{r}, t)g(\mathbf{r}, t) . \qquad (4.32)$$

The neuron activation signal, $q(\mathbf{r}, t)$, is obtained through the relation

$$q(\mathbf{r}, t) = g(\mathbf{r}, t)[K(\mathbf{r}, t) * s(\mathbf{r}, t)] . \qquad (4.33)$$

The CGC loop includes a low-pass temporal filter that integrates the neuron activation signal, which usually has the following impulse response

$$v(\mathbf{r}, t) = B\, q(\mathbf{r}, t) * [\mathrm{u}(t)\, e^{\frac{-t}{\tau}}] , \qquad (4.34)$$

where parameter B controls the amplitude and τ the time duration of the integration. Finally, the signal $q(\mathbf{r}, t)$ passes through a static nonlinear function, resulting in a factor that modulates the RF output. The nonlinearity function has the form

$$g(\mathbf{r}, t) = \frac{1}{1 + \{[v(\mathbf{r}, t)]_+\}^4} , \qquad (4.35)$$

where $[x]_+ = x\, \mathrm{u}(x)$ is the rectification operator.

At the output of the model, the activation signal is rectified to obtain an instantaneous firing rate of the RGC, $r(\mathbf{r}, t)$. The rectification function has the form

$$r(\mathbf{r}, t) = \tilde{\alpha}[q(\mathbf{r}, t) + \Theta]_+ , \qquad (4.36)$$

where $\tilde{\alpha}$ establishes the scale and Θ the baseline for the firing rate. When the stimulus $s(\mathbf{r}, t)$ is spatially uniform and temporally constant, the activity signal $q(\mathbf{r}, t)$ is zero. Thus, the neuron response is just the baseline neural activity, which is equal to the firing rate, $r(\mathbf{r}, t) = \tilde{\alpha}\Theta$.

Computational Implementation The spatial processing carried out by the RF, corresponding to the first block of the model depicted in Fig. 4.6 inside the RF outer block, corresponds to the difference of two-dimensional Gaussian functions as in Eq. (4.30). This kernel can be obtained as the product of two one-dimensional Gaussian bell-shaped curves for the center, one Gaussian for each direction, summed with the product of two other one-dimensional Gaussian curves for the outer sub-kernel. Taking the spatial center sub-kernel, $K_{s_C}(x, y)$ as an example, this can be written as

$$K_{s_C}(x, y) = \frac{1}{2\pi\sigma_{C_x}\sigma_{C_y}} e^{-\frac{1}{2}\left(\frac{x^2}{\sigma_{C_x}^2} + \frac{y^2}{\sigma_{C_y}^2}\right)} = \frac{1}{2\pi\sigma_C^2} e^{-\frac{1}{2}\frac{x^2+y^2}{\sigma_C^2}}, \qquad (4.37)$$

where $x, y \in \mathbb{R}$. The last equality in Eq. (4.37) results from the fact that the RF has no preferred direction, so $\sigma_{C_x} = \sigma_{C_y} = \sigma_C$. The outer sub-kernel has an identical form. Thus, applying different weights to the center and outer sub-kernels the expression in Eq. (4.30) is obtained, where we used the fact that the squared magnitude of the position vector, $\mathbf{r} = [x \ y]^T$, is given by $\mathbf{r}^2 = \mathbf{r}^T\mathbf{r} = x^2 + y^2$.

In order to implement the analog filter given by Eq. (4.30) in a digital computer, it must first be converted into the discrete form. To discretize Eq. (4.30), we start by making $x = n_1 \Delta x$ and $y = n_2 \Delta y$, where Δx and Δy are the spatial sampling interval length in the xx and yy directions, respectively, and $n_1, n_2 \in \mathbb{Z}$. Usually the sampling grid is equally spaced in the xx and yy directions, so that $\Delta x = \Delta y = \Delta r$. Starting by substituting the expression for sampling x and y in Eq. (4.37), we get

$$K_{s_C}(n_1\Delta r, n_2\Delta r) = C\, e^{-\frac{\Delta r^2}{2\sigma_C^2}(n_1^2+n_2^2)}, \qquad (4.38)$$

where C is a constant to be determined. To obtain the correct value for the constant C, one must remember that the multiplicative factors in Eq. (4.37) were included in order to normalize the integral value of the Gaussian function to one. After sampling the Gaussian function to get a total sum equal to one, it must be normalized by the width of the sampling interval, so Eq. (4.38) becomes

$$K_{s_C}(n_1\Delta r, n_2\Delta r) = \frac{\Delta r^2}{2\pi\sigma_C^2} e^{-\frac{\Delta r^2}{2\sigma_C^2}(n_1^2+n_2^2)}. \quad (4.39)$$

Finally, introducing two terms like the one in Eq. (4.39) for the center and the other for the outer components of the RF, with weights A_C and A_S, respectively, and by omitting the argument dependence on Δr so that $K_s(n_1\Delta r, n_2\Delta r) = K_s[n_1, n_2]$, the discrete equation for the spatial DoG kernel becomes

$$K_s[n_1, n_2] = \frac{A_C}{2\pi(\sigma_C/\Delta r)^2} e^{-\frac{\Delta r^2}{2\sigma_C^2}(n_1^2+n_2^2)} - \frac{A_S}{2\pi(\sigma_S/\Delta r)^2} e^{-\frac{\Delta r^2}{2\sigma_S^2}(n_1^2+n_2^2)}. \quad (4.40)$$

The operation of the spatial kernel on the stimulus corresponds to its convolution with the two-dimensional matrix of the filter, for each frame obtained at each instant of time sampling [Lim (1990)]. The shape of this spatial kernel is similar to the one of the LoG two-dimensional function, proposed for detecting intensity changes or contours in an image [Lim (1990); Wulf (2001)]. Figure 4.7 shows the contour plot of a DoG, from Eq. (4.40), with $\sigma_C = 80$ μm, $\sigma_S = 3\sigma_C$, $A_C = 3$, $A_S = 0.8A_C$, for a spatial sampling period of $\Delta r = 1$ μm.

The second block of the RF, corresponding to the temporal kernel, expression Eq. (4.31) in continuous time, has the Laplace transform [Oppenheim et al. (1999b)]

$$K_t(s) = \frac{s}{\alpha + s}, \quad (4.41)$$

where s is the Laplace complex variable. The transfer function of Eq. (4.41) has a pole at $s = -\alpha$ and a zero at $s = 0$, corresponding to a high-pass temporal filter.

To achieve the discretization of this high-pass filter, we can map the Laplace frequency domain into the discrete z-transform domain by means of the bilinear transform. The bilinear transform is an adequate frequency mapping used to transform a continuous-time system representation (see Exercise 4.6), in the s-domain into the discrete-time z-domain [Oppenheim et al. (1999a)] which has the expression

$$s = \frac{2}{T_s}\frac{1-z^{-1}}{1+z^{-1}}, \quad (4.42)$$

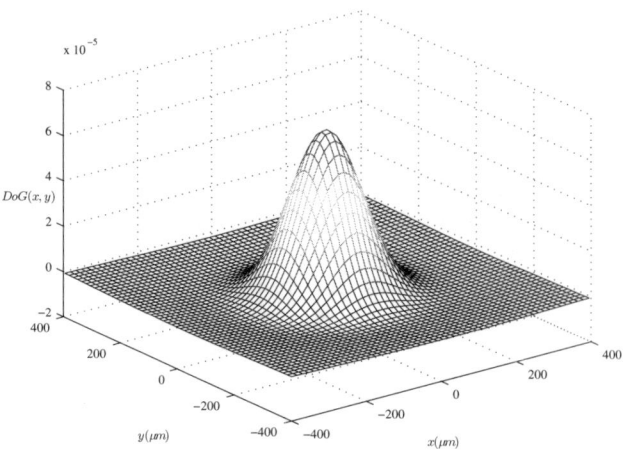

Fig. 4.7 Discrete spatial DoG.

where T_s is the time sampling period.

By applying the bilinear transform to Eq. (4.41) we get

$$K_t(z) = K_t(s)\big|_{s=\frac{2}{T_s}\frac{1-z^{-1}}{1+z^{-1}}}$$
$$= \frac{2}{2+\alpha T_s} \frac{1-z^{-1}}{1-\left(\frac{2-\alpha T_s}{2+\alpha T_s}\right)z^{-1}} . \qquad (4.43)$$

This digital filter has a pole at $z = (2 - \alpha T_s)/(2 + \alpha T_s)$, inside the unit circle.

Applying the properties of the z-transform [Ziemer et al. (1998)], the difference equation relating the input and the output of the temporal kernel, designated by $x[n]$ and $c[n]$ in Fig. 4.6, respectively, can be written in a suitable form to be implemented in a computer as:

$$c[n] = \frac{2 - \alpha T_s}{2 + \alpha T_s} c[n-1] + \frac{2}{2 + \alpha T_s}(x[n] - x[n-1]) . \qquad (4.44)$$

The CGC is composed of two blocks: a low-pass filter and a nonlinear function. The low-pass filter of the CGC can be converted into an equivalent digital filter following the same procedure used before for the temporal kernel.

The Laplace transform of the CGC low-pass filter takes the form:

$$h(t) = B\ \mathrm{u}(t)\exp(-t/\tau) \xrightarrow{\mathcal{L}} H(s) = \frac{B}{1/\tau + s} , \qquad (4.45)$$

and by applying the bilinear transform, the expression for its transfer function is

$$H(z) = H(s)\big|_{s=\frac{2}{T_s}\frac{1-z^{-1}}{1+z^{-1}}}$$

$$= \frac{B\tau T_s}{2\tau + T_s} \frac{1+z^{-1}}{1-\left(\frac{2\tau-T_s}{2\tau+T_s}\right)z^{-1}} . \qquad (4.46)$$

By using the z-transform properties, we get the difference equation relating the output signal $v[n]$ with the input signal $q[n]$ of the CGC low-pass filter:

$$v[n] = \frac{2\tau - T_s}{2\tau + T_s} v[n-1] + \frac{B\tau T_s}{2\tau + T_s}(q[n] + q[n-1]) . \qquad (4.47)$$

The nonlinearity of the model described by Eq. (4.35) can be directly implemented in the digital domain with no difficulty. Finally, the last block of the model, denoted by the rectifier block in the model structure of Fig. 4.6, corresponds to Eq. (4.36), whose computational implementation is also straightforward.

This model produces at its output the RGC firing rate. Thus, to obtain the neural response, a spike generator that generates spikes according to the input firing rate is required (see Sec. 3.4).

4.4.2 Stochastic Models

The stochastic retina model discussed in this section was proposed in the reference [Keat *et al.* (2001)]. It produces as its output the spike train itself, avoiding the need to use an external spike generator. The produced spike trains change from trial to trial for the same stimulus due to the intrinsic stochastic behavior of the model. It attempts to directly model the variability present in the responses of the RGCs.

While other models take into account that these variabilities are part of a stochastic process adopted to generate the spike trains (the process is usually considered to be a Poisson process), this model outputs the spike train, directly modeling its variability by several means along the stimulus processing. The effort in the development of the model to include the variability of the spike trains, and not leave it for the spike generator, is justified by its proponents considering the fact that the variability of the spike trains are kept within a limited interval, smaller than that of a Poisson process, and not accurately accounted for by an external spike generator. The variability exhibited by the spike trains is modeled by the inclusion of two Gaussian noise sources within the model.

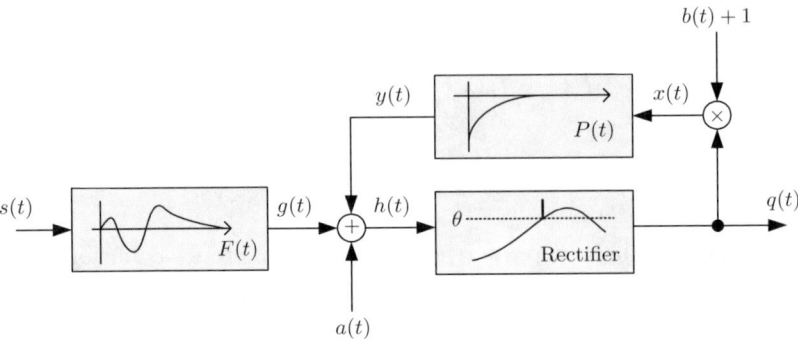

Fig. 4.8   Block diagram of the pseudo-stochastic model.

This model is only concerned with the temporal dependencies of the retina response with the stimulus. However, it is suitable to be extended in order to include a spatial treatment of the stimulus. In particular, if it is assumed that the temporal and spatial processing of the stimulus can be separated, as discussed for the deterministic model, the spatial processing can be modeled by an adequate function [Dayan and Abbot (2001)], such as a DoG [Wandell (1995)].

This model was originally applied to predict the response of the retina to Gaussian random flicker stimuli for several different types of RGCs from different vertebrates, namely rabbit, salamander, and cat. It was also applied to model LGN neuron cells in the case of the cat [Keat et al. (2001)].

Model Description   The structure of the stochastic model is represented in the block diagram of Fig. 4.8. The temporal stimulus $s(t)$ is filtered by a linear filter, with impulse response $F(t)$, producing the generator potential $g(t)$. The filter function $F(t)$ is synthesized using a linear combination of orthonormal functions, f_j, weighted by k_j:

$$F(t) = \sum_{j=1}^{N} k_j f_j(t) \ . \tag{4.48}$$

The basis functions are distorted sinusoidal functions with expression:

$$f_j(t) = \begin{cases} \sin\left(\pi j \left(2\frac{t}{\tau_F} - \left(\frac{t}{\tau_F}\right)^2\right)\right) & \text{if } 0 \leq t \leq \tau_F \\ 0 & \text{otherwise} \end{cases} \ . \tag{4.49}$$

These basis functions, represented in Fig. 4.9, are preferred because of the lower number of parameters needed to reproduce the filter waveform accurately regarding, for example, the traditional sine functions, as in a Fourier series. This filter waveform, which is similar to the STA, has an impulse response with a high amplitude near the origin that decays rapidly over longer times. The parameter τ_F controls the time duration of the filter response.

The signal $g(t)$ in Fig. 4.8 results from the convolution of the stimulus $s(t)$ with this linear filter $F(t)$:

$$g(t) = \int_{-\infty}^{t} s(\tau) F(t-\tau) d\tau \ . \tag{4.50}$$

Fifteen different components with the form of Eq. (4.49) are used to synthesize $F(t)$, which corresponds to making $N = 15$ in Eq. (4.48). This filter selects the stimuli patterns to which the model will fire events, because the signal $g(t)$ will be strongest when the visual stimulus follows a pattern similar to this filter response. Most of the model's parameters are used to appropriately adjust this filter function.

Posteriorly, the signal $g(t)$ is summed with a noise component $a(t)$, and with a feedback signal coming from the feedback block, resulting in the signal $h(t)$, which is then compared with a threshold. The threshold block is composed of three main parts: i) the signal $h(t)$ is compared with a threshold level θ, so that it has a term of the form $\delta(h(t) - \theta)$, in order to guarantee the model only fires one spike when the signal crosses the threshold in the upward direction; ii) it has a term of the form $\mathrm{u}(\dot{h}(t))$, where $\mathrm{u}(t)$ is the continuous Heaviside unit step function; and finally, iii) these two terms are multiplied by the derivative of $h(t)$ with respect to time, denoted by $\dot{h}(t)$. Thus, only when the signal crosses the threshold from below is a spike fired, while when the signal $h(t)$ crosses the threshold from above nothing happens, and the firing intensity is proportional to the intensity increase of $h(t)$.

These first two blocks, the filter $F(t)$ and the threshold block, are intended to predict the time occurrence of the firing events, correspondingly making $a(t) = b(t) = P(t) = 0$ in the model structure of Fig. 4.8. In order to predict the correct number of firing events, a feedback block is introduced in the model to take into account the refractory period of the ganglion cells after a firing event. Each fired spike triggers a negative afterpotential $P(t)$ that is added to the generator potential $g(t)$, lowering the signal $h(t)$ immediately after a firing. The after-potential function $P(t)$ has the form

$$P(t) = B \ \mathrm{u}(t) e^{-t/\tau_p} \ , \tag{4.51}$$

where B and τ_p are two other parameters of the model that define the characteristics of the low-pass filter. The parameter B controls the amplitude of the after potential, while τ_p gives the time decay of the after potential.

The after-potential makes the signal $h(t)$ drop below the threshold after the firing of a spike. However, if $g(t)$ continues to rise in such a way that it compensates for the negative potential $P(t)$, the model will fire again, such that large increments of $g(t)$ lead to a train of several spikes. After a firing event, the signal $h(t)$ is lower than $g(t)$ as a result of the accumulated after potentials, and the probability of subsequent firing events is reduced until the after-potential decays. This negative feedback loop serves to simulate both repetitive firing within a firing event and to include the refractory period after a firing.

The input signal of the feedback block can be written as:

$$q(t) = \delta(h(t) - \theta)\dot{h}(t)\,\mathrm{u}(\dot{h}(t))\,, \tag{4.52}$$

where $\dot{h}(t)$ is included in the threshold function such that the potential block input in the feedback loop is proportional to the slope, or instantaneous increase, of $h(t)$. With the addition of the feedback loop, the generator potential becomes

$$h(t) = g(t) + a(t) + \int_{-\infty}^{t} q(\tau)(1 + b(\tau))P(t-\tau)d\tau\,. \tag{4.53}$$

The output spike train is a series of delta functions, as described by Eq. (3.7), occurring at time instants t_i, whenever the generator potential $h(t)$ crosses the threshold θ from below. Its expression is equal to Eq. (4.52) without the $h(t)$ derivative term, and is written as

$$\hat{\rho}(t) = \delta(h(t) - \theta)\,\mathrm{u}(\dot{h}(t))\,. \tag{4.54}$$

As mentioned previously, this model also attempts to model the variability of the neural response from trial to trial. This variability comprises the variation in the total number of spikes and the variation in the time instants that spikes occur. This is modeled by including two Gaussian noise sources: $a(t)$ and $b(t)$. The amplitudes of the noise signals change over time, and consequently, the spike trains change between trials.

The random signal $a(t)$ is added to the generator potential $g(t)$ before the threshold block, introducing a random variability into the exact time of occurrence of the threshold crossing. This noise source has a Gaussian distribution with a zero mean, a standard deviation of σ_a and an exponentially decaying autocorrelation function with time constant τ_a. The

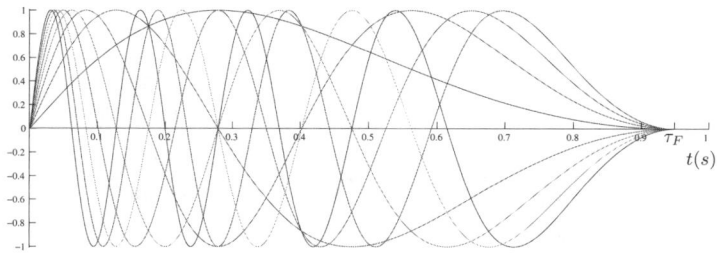

Fig. 4.9 Distorted sinus basis functions ($\tau_F = 0.95s$).

variability introduced by $a(t)$ was not sufficient to reproduce the neural response variability, despite the fact that the spike occurrence variability was well modeled. The variability in the spike number was lower than the observed results from real neurons, and therefore the noise source $b(t)$ was also included.

The noise source $b(t)$ drives, in conjunction with the output of the threshold block, the negative after-potential generator by randomly modulating its amplitude after each spike. This noise source has a Gaussian distribution with zero mean and standard deviation σ_b.

Computational Implementation The stochastic retina model described in the previous section considers continuous signals and systems, and must be discretized in order to be implemented in a digital computer. The steps to implement this retina model in a digital computer are described.

The impulse response $F(t)$ of the filter, represented by the first block in Fig. 4.8, was decomposed as a linear combination of basis functions consisting of distorted sinus functions described by Eq. (4.49). These functions were discretized using a sampling period T_s, for a filter length equal to τ_F, for $j = 1, \ldots, N$. After being sampled, the resulting N vectors were orthonormalized using the Gram-Schmidt procedure [Arfken and Weber (2005)], leading to the vectors $\mathbf{f}_j$, where $j = 1, \ldots, N$. Figure 4.9 shows a plot of the first eight of these distorted sinus functions, with $j = 1, \ldots, 8$.

The filter impulse response $F(t)$ was initialized with a shape equal to the time reverse of the STA. The STA was defined in Section 3.2.3 in its continuous and discrete form, and it is denoted in its discrete-time form by the vector $\mathbf{s}_{spk}$. The time reverse of the STA was decomposed as a

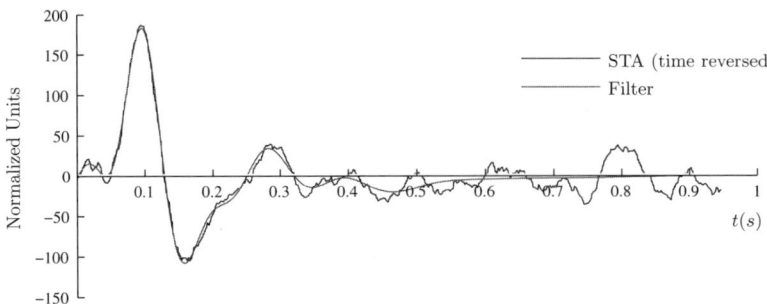

Fig. 4.10 Time reverse of the spike triggered average and its representation by bases functions ($\tau_F = 0.95s; T_s = 1$ ms).

linear combination of the basis functions, as Eq. (4.48) indicates. The discrete orthonormal vectors $\mathbf{f}_j$, with $j = 1, \ldots, N$, were brought together to compose the matrix $\mathbf{F}_\perp$, which is written as

$$\mathbf{F}_\perp = \begin{bmatrix} | & | & & | \\ \mathbf{f}_1 & \mathbf{f}_2 & \cdots & \mathbf{f}_N \\ | & | & & | \end{bmatrix} . \quad (4.55)$$

The coefficients k_j, $j = 1, \ldots, N$ in Eq. (4.48) that are components of the vector $\mathbf{k}$ were initialized with the values resulting from the matrix-vector product:

$$\mathbf{k} = \mathbf{F}_\perp^T \tilde{\mathbf{s}}_{spk} , \quad (4.56)$$

where $\tilde{\mathbf{s}}_{spk}$ is the vectorial representation of the time reverse of the STA.

Figure 4.10 shows the time reverse of the STA superimposed with its reconstruction using the basis functions of Fig. 4.9 after orthonormalization. The filter impulse response, represented by the vector $\mathbf{F}$, is obtained through the matrix product

$$\mathbf{F} = \mathbf{F}_\perp \mathbf{k} . \quad (4.57)$$

The filter impulse response vector, $\mathbf{F}$, was used to filter the stimulus entering the model.

The threshold block in Fig. 4.8, corresponding to Eq. (4.52), has a direct implementation. If the discrete input signal $h[n-1]$ is smaller than the threshold θ, but $h[n]$ is bigger than θ (meaning that there is a threshold crossing in the upward direction), a spike is fired and the output signal has the value $q[n] = h[n] - h[n-1]$. If this situation does not occur, then the output signal is made equal to $q[n] = 0$. It is worth noting that the signal $q[n]$ is proportional to the slope of $h[n]$ whenever it crosses the threshold from below, which means that the feedback potential block is excited by a signal proportional to the slope of the generator potential. The signal $q[n]$ can be computed by using the discrete counterpart of Eq. (4.52),

$$q[n] = \delta\left[h[n] - \theta\right] \mathrm{u}\left[h[n] - h[n-1]\right](h[n] - h[n-1]) \ . \tag{4.58}$$

As a matter of fact, in order to compute the equation Eq. (4.58), the term $\delta[h[n] - \theta]$ must be made equal to one whenever $h[n]$ is in the vicinity of θ, that is when $|h[n] - \theta| < \gamma$, since the signal $h[n]$ would be exactly equal to θ only by chance.

The signal $q[n]$ is convolved with the negative potential that has the continuous expression Eq. (4.51). This potential corresponds to a low-pass continuous filter, as can be seen from its Laplace transform in Eq. (4.59). The discrete counterpart of the after-potential can be obtained by applying the bilinear transform to map the Laplace frequency domain into the z-transform domain.

The Laplace transform of the negative potential is

$$P(t) = B\,\mathrm{u}(t)e^{-t/\tau_p} \xrightarrow{\mathcal{L}} P(s) = \frac{B}{1/\tau_p + s}\ , \tag{4.59}$$

which has a unique pole at $s = -1/\tau_p$ corresponding to a stable system. The application of the bilinear transform to Eq. (4.59) leads to the same equivalent discrete filter in the z-transform domain as obtained in Eq. (4.47) with $\tau = \tau_p$:

$$\begin{aligned} P(z) &= P(s)\big|_{s=\frac{2}{T_s}\frac{1-z^{-1}}{1+z^{-1}}} \\ &= \frac{B\tau_p T_s}{2\tau_p + T_s}\frac{1+z^{-1}}{1-(\frac{2\tau_p-T_s}{2\tau_p+T_s})z^{-1}}\ . \end{aligned} \tag{4.60}$$

Referring to the properties of the z-transform [Ifeachor and Jervis (2002)], the difference equation for the filter corresponding to the negative feedback potential obtained from Eq. (4.60) is

$$y[n] = \frac{2\tau_p - T_s}{2\tau_p + T_s}y[n-1] + \frac{B\tau_p T_s}{2\tau_p + T_s}(x[n] + x[n-1])\ , \tag{4.61}$$

which is used in the computational implementation of the model.

The last items to include in the discrete implementation of the stochastic model are the Gaussian noise sources: $a(t)$ and $b(t)$. Concerning $b(t)$, it is enough to generate a random sequence with a Gaussian distributed amplitude with a zero mean and unit variance, and multiply each new generated sample by the desired standard deviation σ_b. In the case of $a(t)$, in addition to the variance σ_a^2, this noise sequence has an autocorrelation function having an exponential form with a constant decay rate equal to τ_a.

To generate a discrete noise sequence equivalent to $a(t)$, a linear system driven by a discrete white noise sequence with a zero mean and unit variance can be used. The transfer function of this system must be determined such that its output is a random sequence with the desired statistical characteristics. This process is called prewhitening, as it is the reverse of whitening [Orfanidis (1990)]. The required continuous noise has an autocorrelation function with the form

$$R_a(\tau) = \sigma_a^2 e^{-|\tau|/\tau_a}, \qquad (4.62)$$

that takes the value $R_a(0) = \sigma_a^2$ for a lag $\tau = 0$, corresponding to the variance of the continuous random noise. The discrete noise sequence should have an autocorrelation sequence equal to Eq. (4.62) at the sampling points. Using a sampling period equal to T_s, the autocorrelation sequence becomes

$$R_a[l] = \sigma_a^2\ e^{-|lT_s|/\tau_a} = \sigma_a^2\ (\underbrace{e^{-T_s/\tau_a}}_{\varrho})^{|l|} \qquad (4.63)$$

$$= \sigma_a^2\ \varrho^{|l|}\ .$$

By considering a linear shift invariant system with impulse response $h[n]$ subjected to the input $x[n]$ that produces the output signal $y[n] = h[n]*x[n]$, if the autocorrelation of the input signal is $R_x[l]$ and the output autocorrelation is $R_y[l]$, the relationship between these two discrete autocorrelation functions is [Therrien (1992)]

$$R_y[l] = h[l] * h[-l] * R_x[l]\ . \qquad (4.64)$$

By taking the z-transform of this expression, we get

$$S_y[z] = H[z]\ H[z^{-1}]\ S_x[z]\ , \qquad (4.65)$$

where $S_x[z]$ and $S_y[z]$ are the z-transforms of the input and output signals, respectively, corresponding to their spectral densities.

A discrete pure white noise sequence, $w[n]$, with a zero mean and a variance σ_w^2 has the autocorrelation function

$$R_w[l] = \sigma_w^2 \delta[l]\ , \qquad (4.66)$$

and for the particular case of a white noise sequence with unit variance, the spectral density is:

$$R_w[l] = \delta[l] \xrightarrow{\mathcal{Z}} S_w(z) = 1\ . \qquad (4.67)$$

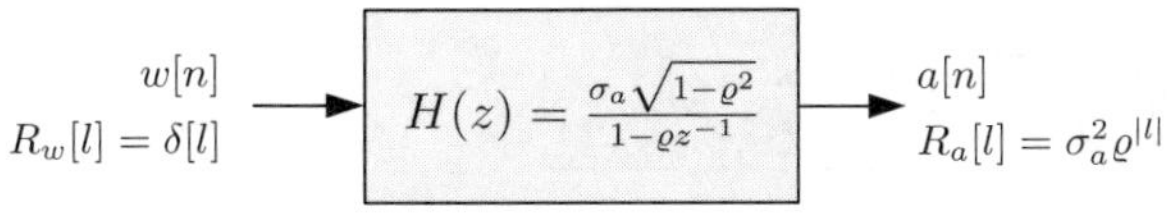

Retina Models

$$w[n] \quad\longrightarrow\quad \boxed{H(z) = \frac{\sigma_a\sqrt{1-\varrho^2}}{1-\varrho z^{-1}}} \quad\longrightarrow\quad a[n]$$
$$R_w[l] = \delta[l] \qquad\qquad\qquad\qquad\qquad R_a[l] = \sigma_a^2 \varrho^{|l|}$$

Fig. 4.11 System function to generate the required noise sequence.

If a discrete white noise sequence with zero mean and unit variance is used as the input for the prewhitening filter, a system transfer function $H(z)$ must be found that transforms this input white noise sequence into a discrete output random sequence with the spectral density

$$R_a[l] = \sigma_a^2 \varrho^{|l|} \xrightarrow{\;z\;} S_a(z) = \frac{\sigma_a^2(1-\varrho^2)}{(1-\varrho z)(1-\varrho z^{-1})}, \qquad |\varrho| < |z| < 1/|\varrho|. \tag{4.68}$$

Plugging the input and output spectral densities, $S_w(z)$ and $S_a(z)$, respectively, into Eq. (4.65), the equation becomes

$$H(z)H(z^{-1}) = \frac{\sigma_a^2(1-\varrho^2)}{(1-\varrho z)(1-\varrho z^{-1})} \tag{4.69}$$

$$= \frac{\sigma_a\sqrt{1-\varrho^2}}{1-\varrho z}\, \frac{\sigma_a\sqrt{1-\varrho^2}}{1-\varrho z^{-1}} \qquad |\varrho| < |z| < 1/|\varrho|.$$

From the factorization in Eq. (4.69), it is possible to identify

$$H(z) = \frac{\sigma_a\sqrt{1-\varrho^2}}{1-\varrho z^{-1}}, \tag{4.70}$$

which corresponds to a causal, stable, and minimum-phase system [Therrien (1992)]. Calculating the inverse z-transform for this transfer function, we get the impulse response

$$h[n] = \sigma_a\sqrt{1-\varrho^2}\,\varrho^n\, \text{u}[n], \tag{4.71}$$

where u$[n]$ is the discrete Heaviside unit step function. By applying the z-transform properties to Eq. (4.70), we obtain the following difference equation for the filter:

$$a[n] = \varrho\, a[n-1] + \sigma_a\sqrt{1-\varrho^2}\, w[n]. \tag{4.72}$$

From Eq. (4.63), we have that $\varrho = e^{-T_s/\tau_a}$ in the last expressions. Figure 4.11 is a sketch of a first order linear system that generates the desired noise sequence when driven by white noise.

The estimation of the discrete neuronal function $\hat{\rho}[n]$ is computed from the input signal of the threshold block $h[n]$ by

$$\hat{\rho}[n] = \delta\bigl[h[n] - \theta\bigr].\,\text{u}\bigl[h[n] - h[n-1]\bigr], \tag{4.73}$$

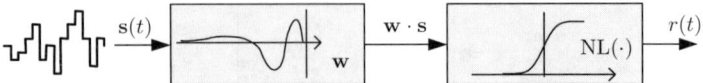

Fig. 4.12 The white noise model structure

which is similar to the expression for the signal $q[n]$ in Eq. (4.58) except for the amplitude. Therefore, we can compute the output of the threshold block from $q[n]$ as:

$$\hat{\rho}[n] = \begin{cases} 1 & q[n] > 0 \\ 0 & q[n] = 0 \end{cases}, \qquad (4.74)$$

that can be written simply as $\hat{\rho}[n] = 1 - \delta[q[n]]$. It should be noted from Eq. (4.58) that $q[n] \geq 0$.

4.4.3 White Noise based Models

A simple model based on a basic form of white noise analysis of the RGCs' responses to random light patterns was proposed in [Chichilnisky (2001)]. White noise analysis has some relevant features; for example, it can generate a quantitative model for the spatial, temporal and spectral responses of the visual system neurons by taking into account its nonlinearities [Chichilnisky (2001); Rust et al. (2004); Simoncelli et al. (2004)].

Model Description The white noise model is portrayed as a block diagram in Fig. 4.12. The first block is a linear filter with an impulse response equal to the time-reverse of vector **w**, which is proportional to the STA if the stimulus space possesses certain characteristics. The second block is a nonlinear function that maps the generator potential signal, given by the inner vector product $\mathbf{w} \cdot \mathbf{s} = \mathbf{w}^T \mathbf{s}$, onto the retina firing rate. The filter impulse response is obtained by white noise analysis of the ganglion cell responses, and the nonlinear function on the second block is obtained by fitting a curve to the generator potential plotted against the firing rate. This retina model gives the neuronal firing rate as the output, assuming that it depends only on the generator signal. The generator signal is a linear combination of the visual stimulus that reaches the retina over a specific region and time period. In order to have a spike train, a spike generator must be used to convert the estimated firing rate to the neural function.

The time interval T, during which the retina is stimulated and the produced spike trains are recorded, is discretized into time bins of width

Δt, resulting in a total recording number of time bins $N = \lfloor T/\Delta t \rfloor$. The number of spikes observed in the time bin n is represented by f_n, where $1 \leq n \leq N$. Furthermore, the number of spikes is related to the neural response function $\rho(t)$, defined in Eq. (3.7), by

$$f_n = \int_{(n-1)\cdot\Delta t}^{n\cdot\Delta t} \rho(\tau)d\tau \ . \qquad (4.75)$$

Each time bin n has a corresponding stimulus vector, which is represented by $\mathbf{s}_n$, with dimension k, whose elements are the stimulus intensities as a function of space and time in the time bins immediately preceding the instant $n \cdot \Delta t$. That is,

$$\mathbf{s}_n = \begin{bmatrix} s(\mathbf{r}, n\Delta t) \\ s(\mathbf{r}, (n-1)\Delta t) \\ \vdots \\ s(\mathbf{r}, (n-k+1)\Delta t) \end{bmatrix} . \qquad (4.76)$$

The definition of $\mathbf{s}_n$ in Eq. (4.76) considers spatial information, which is indicated by the displacement vector $\mathbf{r}$ in its elements, since the stimulus can carry spatial information. For the case when the stimulus is spatially uniform, this dependency is dropped. The displacement vector $\mathbf{r}$ spans the retinal area that influences the RGC under analysis, corresponding to their RF. In general, it is a matrix of positions for each value of the time index n.

The time duration of the stimulus vector $\mathbf{s}_n$, which is equal to $k\Delta t$, must be sufficiently long to exceed the ganglion cell's memory. This corresponds to the period over which the stimulus can affect the cell response; thus, the neuron response f_n at time $n\Delta t$ depends only on the stimulus $\mathbf{s}_n$.

Each stimulus vector $\mathbf{s}_n$ can be viewed as a point in a k-dimensional stimulus space S, and it is assumed that they are drawn randomly from S with a probability distribution given by $P(\mathbf{s})$. This probability distribution is considered to be radially symmetric about the origin in stimulus space. Thus, any two stimulus vectors $\mathbf{s}$ and $\mathbf{s}^* \in S$ with equal vector length have equal probability of being drawn from the distribution, that is,

$$|\mathbf{s}| = |\mathbf{s}^*| \Rightarrow P(\mathbf{s}) = P(\mathbf{s}^*) \ . \qquad (4.77)$$

The radial symmetry implies negative entries in some of the stimulus vectors, meaning that the stimulus entries specify the contrast, or deviation, from a mean intensity level. These stimulus vectors can be generated using a Gaussian white noise sequence, which corresponds to sampling a Gaussian density function with zero mean and a standard deviation equal to the desired contrast. In general, the stimulus intensity for every spatial location, time bin, and wavelength could be drawn from Gaussian noise sequences.

The modeled neural response, $R(\mathbf{s})$, is the average value of the spike count f in the time bin immediately following the stimulus $\mathbf{s}$. This can be written as

$$R(\mathbf{s}) = \langle f | \mathbf{s} \rangle = \sum_f f P(f|\mathbf{s}) , \qquad (4.78)$$

which states that $R(\mathbf{s})$ is proportional to the expected response given the stimulus. In terms of the cell firing rate $r(t)$, the relation is

$$R(\mathbf{s}_n) = r(n\Delta t) \cdot \Delta t . \qquad (4.79)$$

The angle bracket notation in Eq. (4.78) represents the trial average across experiments for the same stimulus. The right hand equality in Eq. (4.78) comes from the definition of statistical expectation, where $P(f|\mathbf{s})$ is the probability distribution of the number of spikes f given a certain stimulus $\mathbf{s}$. To be successful, the final model should predict the average number of spikes per time bin observed after the presentation of a given stimulus.

The white noise analysis estimates $R(\mathbf{s})$. For the model in analysis, it is assumed that $R(\mathbf{s})$ is a static nonlinear functional of a real linear function of the stimulus:

$$R(\mathbf{s}) = \text{NL}(\mathbf{w} \cdot \mathbf{s}) , \qquad (4.80)$$

where $\mathbf{w}$ is a fixed weighting vector and $\text{NL}(\cdot)$ is a real-valued nonlinear function of one variable. The vector $\mathbf{w}$ weights the neuron stimulus intensities over space and time and defines the neuron stimulus selectivity. The dot product $\mathbf{w} \cdot \mathbf{s} = \mathbf{w}^T \mathbf{s}$ is the generator signal that controls the firing rate through the nonlinear function $\text{NL}(\cdot)$. (See Fig. 4.12.)

Equation (4.80) shows that the model response does not depend on previous responses, but solely on the stimulus, meaning that the spikes are generated by a Poisson-like process with a rate parameter equal to the expected response $R(\mathbf{s})$. To obtain a spike train, a spike generator must be added at the output of the model in Fig. 4.12.

The weight vector $\mathbf{w}$, as in many retinal ganglion cells models, is related to the STA. For this particular setup, the STA (see Sec. 3.2.3), that corresponds to the average stimulus preceding the spikes fired by the neuron, is given by

$$\mathbf{s}_{spk} = \frac{\sum_{n=1}^{N} \mathbf{s}_n f_n}{\sum_{n=1}^{N} f_n}, \qquad (4.81)$$

where N again denotes the total number of time bins. Equation (4.81) represents the discrete spike triggered average $\mathbf{s}_{spk}$, which is now a vector of dimension k (the same dimension as $\mathbf{s}_n$); the time dependency of the definition given by Eq. (3.39) is implicit.

Dividing the numerator and denominator of Eq. (4.81) by the total time duration T of the stimulus, it becomes

$$\mathbf{s}_{spk} = \frac{\frac{1}{T}\sum_{n=1}^{N} \mathbf{s}_n f_n}{\frac{1}{T}\sum_{n=1}^{N} f_n}. \qquad (4.82)$$

If the response record lasts for a long time, such that $T \to \infty$, the denominator in Eq. (4.82) is the average firing rate $\langle r \rangle$, defined in Eq. (3.23). Also, as $T \to \infty$ the numerator in Eq. (4.82) tends to the time average $\langle \mathbf{s}f \rangle$. Statistically, this expectation can be expressed as the sum of all stimulus response pairs weighted by the probability of observing that particular stimulus response pair:

$$\langle \mathbf{s}f \rangle = \sum_{\mathbf{s}} \sum_{f} \mathbf{s} f P(\mathbf{s}, f). \qquad (4.83)$$

Using Bayes' rule, and by applying Eq. (4.78), the expression in Eq. (4.83) can be written as

$$\begin{aligned}
\langle \mathbf{s}f \rangle &= \sum_{\mathbf{s}} \sum_{f} \mathbf{s} f P(\mathbf{s}) P(f|\mathbf{s}) \\
&= \sum_{\mathbf{s}} \mathbf{s} P(\mathbf{s}) \underbrace{\sum_{f} f P(f|\mathbf{s})}_{R(\mathbf{s})} \\
&= \sum_{\mathbf{s}} \mathbf{s} P(\mathbf{s}) R(\mathbf{s})
\end{aligned} \qquad (4.84)$$

Replacing the result of Eq. (4.84) in Eq. (4.82) and considering large values of T, the following expression is obtained for the STA:

$$\mathbf{s}_{spk} = \frac{1}{\langle r \rangle} \sum_{\mathbf{s}} \mathbf{s} P(\mathbf{s}) R(\mathbf{s}). \qquad (4.85)$$

The spike triggered average in Eq. (4.85) approaches a sum of stimulus vectors, where each vector is weighted by its probability of being drawn, times the average response it induces, that is normalized by the average firing rate.

The radial symmetry of the stimulus space means that there are two stimulus vectors, $\mathbf{s}^* \in S$ and $\mathbf{s} \in S$, positioned symmetrically relative to the vector $\mathbf{w}$, that have equal probability of being drawn. This means that $P(\mathbf{s}) = P(\mathbf{s}^*)$. Introducing this result into Eq. (4.85), and using the equality in Eq. (4.80), we obtain

$$\begin{aligned}\mathbf{s}_{spk} &= \frac{1}{\langle r \rangle} \sum_{\mathbf{s},\mathbf{s}^*} [\mathbf{s} P(\mathbf{s}) \, \mathrm{NL}(\mathbf{w} \cdot \mathbf{s}) + \mathbf{s}^* P(\mathbf{s}^*) \, \mathrm{NL}(\mathbf{w} \cdot \mathbf{s}^*)] \\ &= \frac{1}{\langle r \rangle} \sum_{\mathbf{s},\mathbf{s}^*} (\mathbf{s}+\mathbf{s}^*) P(\mathbf{s}) \, \mathrm{NL}(\mathbf{w} \cdot \mathbf{s}) \,, \end{aligned} \qquad (4.86)$$

where the factorization in the last equality results from the fact that $\mathbf{s}$ and $\mathbf{s}^*$ have equal probability of being drawn and are symmetric around $\mathbf{w}$. This implies that $\mathbf{w} \cdot \mathbf{s} = \mathbf{w} \cdot \mathbf{s}^*$. The symmetry of $\mathbf{s}$ and $\mathbf{s}^*$ around the vector $\mathbf{w}$ also implies that $\mathbf{s} + \mathbf{s}^*$ is proportional to $\mathbf{w}$. Furthermore, since all other quantities in Eq. (4.86) are scalars, it can be concluded that the vector $\mathbf{w}$ is proportional to the STA:

$$\mathbf{w} \propto \mathbf{s}_{spk} \,. \qquad (4.87)$$

This reasoning leads to the conclusion that the linear part of the model in Fig. 4.12, corresponding to $\mathbf{w}$, is equal to the STA vector $\mathbf{s}_{spk}$ multiplied by a gain factor.

The linear weighting vector $\mathbf{w}$ expresses the way the neuron integrates the visual stimuli. The temporal structure of $\mathbf{w}$ corresponds to the time-reverse of the neuron's impulse response, and the spatial structure of $\mathbf{w}$ describes the neuron's RF. The neuron memory can also be obtained by examining the time duration of the impulse response, and the spectral response of the neuron can be characterized by analyzing the chromatic structure of $\mathbf{w}$.

The gain factor can be taken into account by including a scale factor to adjust the nonlinearity function. Thus, a proportionality constant equal to unity can be used in Eq. (4.87), such that $\mathbf{w} = \mathbf{s}_{spk}$.

Finally, we have to estimate the nonlinear function. Since $\mathrm{NL}(\cdot)$ is a real-valued, positive function of the generator potential $g_n = \mathbf{w} \cdot \mathbf{s}_n$, and the values of f_n are available, it is possible to plot the values of the generator potential g_n against the number of spikes in each bin f_n for every $0 < n \leq N$

in order to estimate the form of NL(·). The nonlinearity can be estimated by plotting the generator potential against the number of spikes it produces and fitting the nonlinear function using a minimization error criterion. For example, the nonlinear function has the expression [Chichilnisky (2001)]:

$$f(x) = r_{max} \, \text{CDF}(\beta x + \gamma), \qquad (4.88)$$

where $\text{CDF}(x)$ represents the cumulative normal density function – the integral of the Gaussian function [Abramowitz and Stegun (1965)]. The parameter r_{max} gives the maximum firing rate of the neuron, β is the sensitivity of the nonlinearity to the generator signal, and γ gives the baseline firing rate, which defines the firing rate in the absence of any visual stimulation. The parameter γ can also be negative, meaning that the generator potential must overcome this value to fire a spike. The drawback of the cumulative density function (CDF), in terms of computational implementation, is that its expression,

$$\text{CDF}(x) = \int_{-\infty}^{x} \frac{1}{\sqrt{2\pi}} e^{-\frac{1}{2}z^2} dz, \qquad (4.89)$$

does not have a closed mathematical form. This imposes some constraints on its implementation in a signal processing system.

Another commonly used nonlinear function is the sigmoidal function [Dayan and Abbot (2001)], with the expression

$$f(x) = \frac{r_{max}}{1 + e^{(g_{1/2} - x)/\Delta g}}, \qquad (4.90)$$

where the parameter r_{max}, once again, represents the neuron maximum firing rate, $g_{1/2}$ is the generator potential value that produces a firing rate with half of its maximum value, and Δg controls how quickly the firing rate increases as a function of the generator potential g. For a negative Δg, the firing rate decreases monotonically with the generator potential. Plots of the STA and of the nonlinear function (using Eq. (4.88) and Eq. (4.90)) for a salamander ON-type and for a rabbit OFF-type brisk RGC cell are displayed in Fig. 4.13. We can notice that both nonlinear functions, Eq. (4.88) and Eq. (4.90), are similar in the interval of interest, which means that both can be used in this model without noticeable differences.

Computational Implementation This simple white noise model can be considered as fitting a curve to the number of fired spikes as a function of a stimulus characteristic, namely the stimulus projection onto the STA,

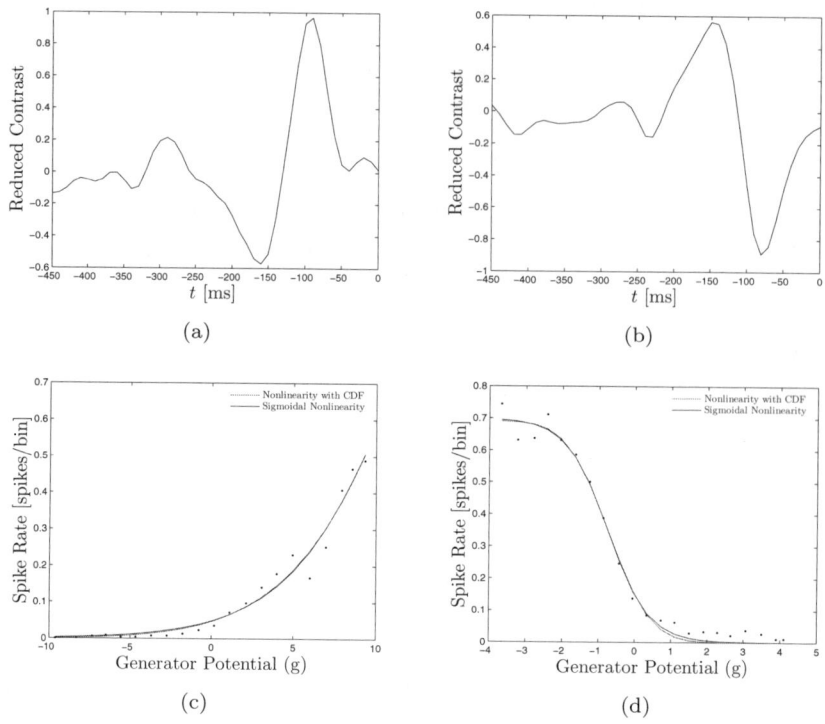

Fig. 4.13 White noise model characterization for a salamander ON-type cell and for a rabbit OFF-type transient brisk cell. The spike triggered average for the salamander cell (a) and the rabbit cell (b). Generator signal versus number of spikes per time bin for the salamander cell (c) and for the rabbit cell (d), with $\Delta t = 10$ ms

which is a common procedure in neuron modeling in theoretical neuroscience [Dayan and Abbot (2001)]. Its implementation in a computer is quite straightforward, since all expressions are already in discrete form.

First, the model parameters must be estimated (the STA) using experimental data; the example in Fig. 4.13 uses a salamander ON-type cell and a rabbit OFF-type brisk transient cell [Keat et al. (2001)]. The vector **w** is obtained from the spike triggered average stimulus, depicted in Fig. 4.13 for both types of cells used. The spike occurrence is considered to be located at the time instant 0, and so the negative time values are the time instants before the spike occurrence.

To fit the nonlinear function we can build a table with the points, corresponding to the generator potential $g_n = \mathbf{w} \cdot \mathbf{s}_n$ and the number of fired spikes f_n for the respective time bins. The number of elements in this table depends on the time bin width. The used time bin, $\Delta t = 10$ ms, was large when compared to the sampling period of $T_s = 1$ ms. The number of spikes per time bin oscillates with the generator potential. The range of the generator potential values was divided into 20 equal subintervals, and the values of the generator potential and of the spikes per bin were averaged for each interval. The plots in Fig. 4.13(c) and Fig. 4.13(d) show the generator potential against the number of spikes per time bin for the analyzed RGC cells. We observe that the nonlinear function shows different behaviors from the plot in Fig. 4.13(c), for the case of the salamander ON-type cell, and in Fig. 4.13(d), for the case of the rabbit OFF-type transient brisk cell. These graphs also show that using the sigmoidal nonlinearity of Eq. (4.90) leads to a curve fit similar to the one obtained with the CDF function within the spread of points in the plot. The sigmoidal nonlinearity is preferred when taking into account computational feasibility.

The dot product of the vector **w** with the succeeding stimulus vectors $\mathbf{s}_n$ can be viewed as a linearly filtered discrete stimulus signal $s[n]$ with an impulse response equal to the time reverse of **w** [Moon and Stirling (2000)]. This approach was followed in the model implementation, and the filter output was used as the argument to drive the nonlinear function. This, in turn, produces the firing rate $r[n]$. In Chap. 5, the results for this model are further described and evaluated.

4.5 Conclusions and Further Reading

In this chapter, we have given a brief overview of the diversity of retina models by picking some representative models from the different classes. Many more examples of models can be found by examining publications related to the topic.

For a thorough list of general structural neuron models you may consult [Gerstner and Kistler (2002)]. Recent references considering white noise-analysis and models of biological systems are [Westwick and Kearney (2003)] and [Marmarelis (2004)]. An example of a statistical analysis procedure applied to retina modeling can be found in [Martins et al. (2007)]. For a more extensive exposition and explanation of the mechanisms responsible for the neuron's dynamics in terms of currents and voltages sources, refer to the book [Dayan and Abbot (2001)].

In addition to the neuron models presented here, the research community is currently attempting to model the responses of cell populations, instead of just a single cell.

It became evident that the tools needed to analyze and implement retina models in a computer are vast, requiring knowledge from scientific computing and signal processing. In addition to the bibliography given in this chapter, we would like to particularly call the attention to the reference [Moon and Stirling (2000)], which presents an extensive set of mathematical methods and signal processing algorithms. For the specific subject of signal processing, there are a great number of good books on the subject – some are classical, like [Oppenheim et al. (1999a)] and [Proakis and Manolakis (2006)]. For the more general subject of signals and systems, we recommend [Oppenheim et al. (1999b)] and [Ziemer et al. (1998)]. For statistical signal processing and stochastic processes, two good references are [Hayes (1996)] and the renewed classic [Papoulis and Pillai (2002)]. For a more practical perspective on signal processing, we suggest [Ifeachor and Jervis (2002)] and [Smith (2003)]. The many tutorials available on the internet should also be considered for a short and informal introduction to the listed subjects.

Exercises

4.1. *With the help of an electric circuit simulator program (e.g., Tina simulator from Texas Instruments), implement the leaky integrate-and-fire neuron model.(Can you further develop the circuit to include a voltage comparator, to compare V_m with V_θ, and to include a switch that short circuits the capacitor whenever V_m crosses the threshold potential V_θ?).*

4.2. *The bilateral Laplace transform of a signal $x(t)$ is defined by*

$$\mathcal{L}\big[x(t)\big] = X(s) = \int_{-\infty}^{+\infty} x(t)\,e^{-st}\,dt\;, \tag{4.91}$$

where the parameter $s = \sigma + j\omega$ is complex. The values of s for which the integral in Eq. (4.91) exist define the region of convergence (ROC) *of the transform.*

Calculate the Laplace transform and the region of convergence (ROC) of the following continuous signals:

4.2.1 *The Dirac delta function, $\delta(t)$, taking into account the properties* in Eq. (3.8) to Eq. (3.10);

4.2.2 $x_1(t) = \mathrm{u}(t)$, of $x_2(t) = \mathrm{u}(-t)$, and of $x_3(t) = \mathrm{u}(t-\tau)$, where $\mathrm{u}(t)$ is the Heaviside step function defined in Eq. (3.31). A signal $x(t)$ for which $x(t) = 0$, $t < 0$ is termed a causal signal. From the results obtained for the ROCs of $x_1(t)$, $x_2(t)$ and $x_3(t)$ can you devise what form should have the ROC of a causal signal?

4.2.3 The high-pass filter of the temporal kernel of the deterministic model given by Eq. (4.31).

4.3. The Laplace transform is particularly useful to solve differential equations, since it transforms a differential equation into a polynomial equation, which is easier to solve.

4.3.1 Show that the Laplace transform is linear, that is:
$$\mathcal{L}\big[a\,x(t) + b\,h(t)\big] = aX(s) + bH(s) \qquad (4.92)$$
where a and b are constant, and $x(t) \xrightarrow{\mathcal{L}} X(s) = \mathcal{L}\big[x(t)\big]$ and $h(t) \xrightarrow{\mathcal{L}} H(s) = \mathcal{L}\big[h(t)\big]$.

4.3.2 Show that the Laplace transform of the first derivative of $x(t)$ is:
$$\frac{dx(t)}{dt} \xrightarrow{\mathcal{L}} sX(s) - x(0) \qquad (4.93)$$
where $X(s) = \mathcal{L}\big[x(t)\big]$, and it is assumed that $x(t) = 0$ for $t < 0$ ($x(t)$ is causal).

4.3.3 Based on the previous exercise, calculate the Laplace transform of the second derivative of $x(t)$. Can you generalize the result for the nth order derivative?

4.3.4 Show that the Laplace transform of the continuous convolution, defined in Eq. (3.25), of two causal signals $y(t) = x(t) * h(t)$ is:
$$\mathcal{L}\big[x(t) * y(t)\big] = X(s)H(s) = Y(s) \qquad (4.94)$$
where $x(t) \xrightarrow{\mathcal{L}} X(s)$, $h(t) \xrightarrow{\mathcal{L}} H(s)$ and $y(t) \xrightarrow{\mathcal{L}} Y(s)$.

4.4. If we have a linear system whose input signal is $x(t)$ and the output signal is $y(t)$ then
$$H(s) = \frac{Y(s)}{X(s)} \qquad (4.95)$$
is termed the continuous transfer function of the system, where $x(t) \xrightarrow{\mathcal{L}} X(s)$, $y(t) \xrightarrow{\mathcal{L}} Y(s)$. And $H(s) = \mathcal{L}\big[h(t)\big]$ is the Laplace transform of the impulse response of the system, $h(t)$.

4.4.1 *Compute the Laplace transform of the differential equation of Eq. (4.11), that models the leaky integrate-and-fire neuron model, and obtain its transfer function. (To obtain the transfer function from Eq. (4.95) consider null initial conditions: $V_m(0) = 0$). With the help of a table of Laplace transform pairs, compute the expression of the impulse response for null initial conditions $V_m(0) = 0$.*

4.4.2 *For a continuous input current, $I_s(t) = I_s \operatorname{u}(t)$, and for an initial membrane at the rest potential, $V_m(0) = V_r$, compute the Laplace transform of the membrane potential for Eq. (4.11). Invert by inspection the resulting Laplace transform to obtain the expression for the membrane potential $V_m(t)$, (compare the result obtained with Eq. (4.15)).*

4.4.3 *Calculate the impulse response for the leaky integrate-and-fire model, modeled by the differential equation Eq. (4.11), with non-null initial conditions ($V_m(0) = V_r$) by taking into account the fact that if $x(t) = \delta(t)$ then, from Eq. (4.95), we have that $H(s) = X(s)$. (This means that when we excite a linear system with the Dirac delta function, $\delta(t)$, the system output is its impulse response, $h(t)$.)*

4.5. *The z-transform is very useful to solve difference equations, since it transforms a difference equation into an algebraic equation. The z-transform of a discrete sequence $x[n]$ is defined as*

$$\mathcal{Z}[x[n]] = X(z) = \sum_{n=-\infty}^{+\infty} x[n] z^{-n}, \quad (4.96)$$

where $z = r\,e^{j\omega}$ is a complex variable. The values of z for which the sum in Eq. (4.96) converges define the region of convergence *(ROC) of the transform in the complex plane.*

Calculate the z-transform and the associated ROC of the following sequences:

4.5.1 $x[n] = a^n \operatorname{u}[n]$.
4.5.2 $x[n] = a^n \operatorname{u}[-n]$
4.5.3 *If $x[n] \xrightarrow{\mathcal{Z}} X(z)$, calculate $\mathcal{Z}[x[-n]]$.*

4.6. *In the computational implementation of the leaky integrate-and-fire model described in Sec. 4.3.2, we used the forward, or Euler, approximation*

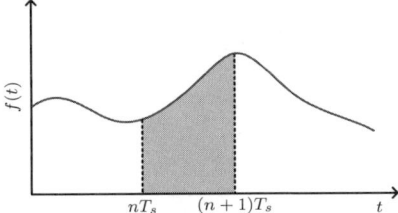

Fig. 4.14 Function integral between nT_s and $(n+1)T_s$.

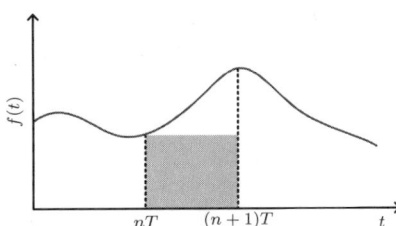

Fig. 4.15 Forward rectangular approximation for the integral.

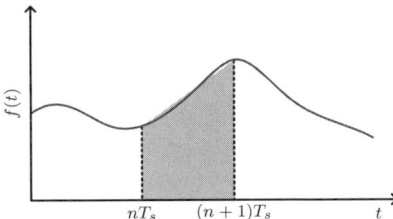

Fig. 4.16 Trapezoidal approximation for the integral.

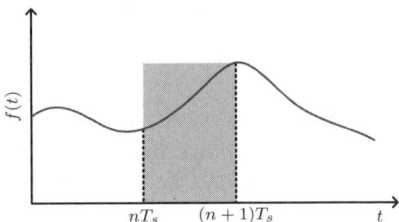

Fig. 4.17 Backward approximation for the integral.

for the first derivative (see Eq. (4.24)) that led to the difference equation of Eq. (4.26).

4.6.1 *(Euler approximation for the integral) The Euler approximation can also be obtained by the forward approximation of the integral:*

$$y(t) = \int_0^t f(\tau)d\tau \,, \qquad (4.97)$$

where the function $f(t)$ is represented in Figure 4.15. If we discretize time with a sampling period T_s, then $y(t)$ can be written for $t = (n+1)T_s$ as

$$\begin{aligned} y((n+1)T_s) &= \int_0^{(n+1)T_s} f(\tau)d\tau \\ &= \underbrace{\int_0^{nT_s} f(\tau)d\tau}_{y(nT_s)} + \int_{nT_s}^{(n+1)T_s} f(\tau)d\tau \,. \end{aligned} \qquad (4.98)$$

If we approximate the second integral in the second member of Eq. (4.98) by the shaded area in Fig. 4.15 so that

$$\int_{nT_s}^{(n+1)T_s} f(\tau)d\tau \simeq f(nT_s)\,T_s\,, \tag{4.99}$$

then the integral in Eq. (4.98) can be approximated by

$$y((n+1)T_s) = y(nT_s) + f(nT_s)\,T_s\,. \tag{4.100}$$

Write Eq. (4.11) in its integral form and apply the approximation for the integral of Eq. (4.100) (where now $y(t)$ is $V_m(t)$ and $f(t) = -V_m(t)/\tau + C_m I_s(t)$) and obtain the difference equation found in Eq. (4.26).

4.6.2 *(Trapezoidal approximation for the integral)* If in Eq. (4.98) we approximate the second integral in the second member by the shaded area in Fig. 4.16 so that

$$\int_{nT_s}^{(n+1)T_s} f(\tau)d\tau \simeq \frac{1}{2}[f(nT_s) + f((n+1)T_s)]\,T_s\,, \tag{4.101}$$

we get that the integral in Eq. (4.98) is approximated by

$$y((n+1)T_s) = y(nT_s) + \frac{1}{2}[f(nT_s) + f((n+1)T_s)]\,T_s\,. \tag{4.102}$$

Obtain the difference equation for the leaky-integrate-and-fire model equation by rewriting Eq. (4.11) in its integral form and by applying the approximation for the integral in Eq. (4.102).

4.6.3 *(Bilinear transform)* Calculate the Laplace transform of Eq. (4.11) and the z-transform of the difference equation obtained in the previous exercise. By comparing the expression obtained in the s-domain (Laplace transform) to that generated in the z-domain (z-transform), show that the correspondence between the two domains can be obtained by making

$$s \to \frac{2}{T_s}\frac{z-1}{z+1}\,. \tag{4.103}$$

The transformation in Eq. (4.103) is known as the bilinear transform (or Tustin's method), and is used to transform a continuous-time system representation in the Laplace domain to a discrete-time representation in the z-transform domain.

4.6.4 *(Backward approximation for the integral)* Repeat the same process for the backward approximation for the integral, displayed in Fig. 4.17, and obtain the respective difference equation and the correspondence between the Laplace and z-transform domain.

4.6.5 Write a program in Matlab to implement the leaky integrate-and-fire model based in the forward approximation and in the trapezoidal approximation for the integral. Run several simulations for different values of the sampling period T_s and for different forms of the stimulus current and compare the results.

4.7. As discussed in Sec. 4.4.1, the spatial receptive field (RF) of a neuron is frequently modeled with a DoG. The central and surrounding width of the Gaussian functions in Eq. (4.30) are related by $\sigma_S^2 = \beta^2 \sigma_C^2$, where $\beta > 1$.

4.7.1 Based on Eq. (4.40), implement the DoG on a computer, and observe the curves for different values of β (particularly for $\beta \approx 1$ and $\beta \gg 1$, and for different values for the weights A_C and A_S).

4.7.2 Pick several types of images, such as natural and urban landscapes, and observe the resulting images after convolving them with the DoG kernels obtained for the extreme and intermediate values of β.

4.8. The frequency response of a system can be obtained by setting $s = j\omega$ in the Laplace transform of its impulse response, which is equal to its Fourier transform in the particular case of the Laplace transform. If $H(j\omega) = H(s)|_{s=j\omega}$, the frequency response in polar form is given by $H(j\omega) = |H(j\omega)| e^{\sphericalangle H(j\omega)}$, where $|H(j\omega)|^2 = H(j\omega) H^*(j\omega)$ ($H^*(j\omega)$ is the complex conjugate of $H(j\omega)$). $|H(j\omega)|$ is the amplitude of the frequency response, and $\sphericalangle H(j\omega)$ is the phase. It is common to plot the amplitude of the frequency response as $20 \log_{10} |H(j\omega)|$, which is expressed in decibels (dB). Plot the amplitude (in dB) and the phase of the frequency response for the temporal RF kernel of the deterministic model using the expression in Eq. (4.31), whose Laplace transform is given by Eq. (4.41). From the frequency amplitude response, can you conclude what type of filter it is (i.e., low-pass, band-pass or high-pass)?

4.9. Show that the z-transform of the sequence

$$x[n] = \alpha a^{|n|} \qquad (4.104)$$

is

$$X(z) = \frac{a(1-a)}{(1-az)(1-az^{-1})}, \qquad (4.105)$$

with an ROC, $|a| < |z| < 1/|a|$. This transform can be obtained directly from the definition in Eq. (4.96) or by employing the results from Exercise 4.5. (This is the same as the z-transform pair of Eq. (4.68)).

4.10. Obtain the discrete equivalent of the RF for the temporal kernel in Eq. (4.31) in the deterministic model by applying the bilinear transform, given by Eq. (4.103), to Eq. (4.41).

4.11. Show that the z-transform of the sequence $x[n - n_0]$, which is $x[n]$ delayed by n_0, is

$$\mathcal{Z}\big[x[n - n_0]\big] = z^{-n_0} X(z), \qquad (4.106)$$

where $x[n] \xrightarrow{\mathcal{Z}} X(z)$. This is called the shifting property and is very useful in signal processing.

4.12. The output $y[n]$ of a discrete linear shift-invariant system, with an impulse response $h[n]$, subjected to the input signal $x[n]$, is given by the convolution of the input with the impulse response $h[n]$:

$$y[n] = x[n] * h[n] = \sum_{k=-\infty}^{+\infty} x[k] h[n-k]. \qquad (4.107)$$

Show that if we have the z-transform pairs: $x[n] \xrightarrow{\mathcal{Z}} X(z)$ and $h[n] \xrightarrow{\mathcal{Z}} H(z)$, the z-transform of the output is

$$y[n] \xrightarrow{\mathcal{Z}} Y(z) = H(z)X(z). \qquad (4.108)$$

The z-transform of the impulse response, $H(z) = Y(z)/X(z)$, is referred to as the system function.

4.13. Consider a linear shift-invariant system that has the impulse response $h[n] = \left(\frac{3}{4}\right)^n u[n]$.

- **4.13.1** Calculate the system output when the input is $x[n] = \left(\frac{3}{4}\right)^n u[n] + (2)^n u[-n - 1]$.
- **4.13.2** Obtain the difference equation relative to this linear system. Implement in Matlab this difference equation and confirm the result obtained in the previous exercise.
- **4.13.3** Calculate the system response to the signal $\delta[n]$. What is its relation to the impulse response?

4.14. *Using the shifting property of the z-transform stated in exercise 4.11, obtain the time domain difference equation in Eq. (4.61) from the z-transform of Eq. (4.60).*

Chapter 5

Neural Activity Metrics and Models Assessment

5.1 Introduction

One of the tasks in the development of a bioelectronic vision system is the choice of a retina model. Therefore, it is imperative to have a measure of the model's performance, and consequently a metric against which the performance is evaluated. Model performance is measured by comparing its output with the responses from a real retina given the same input, by applying a metric.

There are several classes of metrics, depending on the adopted perspective on the neural code. After introducing the concept of a metric in this chapter, we describe several metrics to compare spike trains, firing rates, and sets of spike trains, and discuss their main characteristics. At the end of the chapter, the metrics are applied to tune and assess the functional retina models presented in the previous chapter, with the goal of comparing their performances.

5.2 The Metric Definition

A metric is a function that maps a pair of points, $\mathbf{x} \in \mathbb{S}$ and $\mathbf{y} \in \mathbb{S}$, of a vectorial space $\mathbb{S}$, to a nonnegative real number, d, representing the distance between those two points: $d : \mathbb{S}^2 \to \mathbb{R}_0^+$. To be regarded as a metric in the mathematical sense, the distance function $d(\mathbf{x}, \mathbf{y})$ must have the following properties $\forall \mathbf{x}, \mathbf{y} \in \mathbb{S}$:

$$d(\mathbf{x}, \mathbf{y}) \geq 0, \quad \text{where} \quad d(\mathbf{x}, \mathbf{y}) = 0 \Rightarrow \mathbf{x} = \mathbf{y} \tag{5.1}$$

$$d(\mathbf{x}, \mathbf{y}) = d(\mathbf{y}, \mathbf{x}) \tag{5.2}$$

$$d(\mathbf{x}, \mathbf{y}) \leq d(\mathbf{x}, \mathbf{z}) + d(\mathbf{z}, \mathbf{y}) \quad \text{(triangle inequality)} \tag{5.3}$$

An example of a metric is the Euclidian distance defined in the $\mathbb{R}^n$ Euclidian space. The Euclidian distance corresponds to the distance between two n-dimensional points, with coordinates $\mathbf{x} = (x_1, x_2, \ldots, x_n)$ and $\mathbf{y} = (y_1, y_2, \ldots, y_n)$, defined by $d : \mathbb{R}^n \times \mathbb{R}^n \to \mathbb{R}_0^+$ as

$$d(\mathbf{x}, \mathbf{y}) = \sqrt{(y_1 - x_1)^2 + (y_2 - x_2)^2 + \cdots + (y_n - x_n)^2} \ .$$

To measure the distance between neural responses, suitable metrics have been devised. In the following, different metrics and error measures that have been proposed to evaluate the performance of neuronal models and to analyze neuronal activity are classified and presented. Some of the analyzed distance measures clearly depart from the straight mathematical definition of a metric since they do not possess some of the properties enumerated in Eq. (5.1)–Eq. (5.3). Despite this, they are all still called metrics.

Depending on the adopted perspective on the neural code, we can organize metrics into three main classes: *i*) the firing rate metrics, which compare firing rates; *ii*) the spike train metrics, which directly compare two spike trains; and *iii*) the spike events metrics, which measure the (dis)similarity between bursts of spikes within two sets of spike trains. These metrics look to the neural code from different perspectives, and each of them assign unique weights to the various characteristics of the neural response. After presenting the classes of metrics, we analyze and compare the metrics when applied to salamander ON-type RGC responses to a white noise stimulus.

5.3 Firing Rate Metrics

The firing rate metrics are particularly suited for the analysis of rate code models since they compare firing rates. In the context of neuron model assessment, the comparison of two firing rates comprises a reference firing rate, which is usually obtained from the responses of a retinal neuron, and an estimated firing rate, which corresponds to the model output. Figure 5.1 illustrates the distance measure between the reference firing rate, $r(t)$, and the estimated firing rate, $\hat{r}(t)$.

5.3.1 *Mean Squared Error*

The *mean squared error* (MSE) is one of the most common metrics used in engineering, and it is also suitable to compare firing rates. The mean

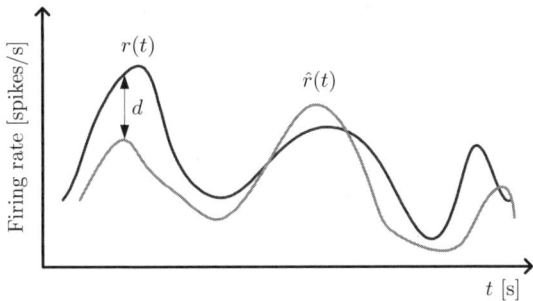

Fig. 5.1 Comparison of two firing rates.

squared error (MSE) between two firing rates is defined by:

$$\mathrm{MSE}(r, \hat{r}) = E\{[r(t) - \hat{r}(t)]^2\} = \frac{1}{T} \int_0^T [r(t) - \hat{r}(t)]^2 dt, \quad (5.4)$$

where $\hat{r}(t)$ is the firing rate that is compared with the reference firing rate $r(t)$ during the interval of time T. The second equality in Eq. (5.4) denotes the expectation operator E; for the case of an ergodic process a time average can be used instead of the ensemble average [Papoulis and Pillai (2002); Therrien (1992)].

For actual data analysis, the firing rate is discretized in time so that the MSE is obtained from

$$\mathrm{MSE}(r, \hat{r}) = \frac{1}{N} \sum_{n=0}^{N-1} (r[n] - \hat{r}[n])^2, \quad (5.5)$$

where N represents the number of time bins into which the spike trains were discretized.

5.3.2 Normalized Mean Squared Error

Another statistic, used to measure how well a given model captures the fast modulations in the observed firing rate, is obtained by dividing the MSE by the variance of the observed firing rate [Berry II and Meister (1998)]. This metric, the *normalized mean squared error* (NMSE), is defined by

$$\mathrm{NMSE}(r, \hat{r}) = \frac{\int_0^T [r(t) - \hat{r}(t)]^2 dt}{\int_0^T [r(t) - \langle r \rangle]^2 dt}. \quad (5.6)$$

When the firing rate is discretized into time bins of finite length, the expression for the normalized mean squared error (NMSE) can be written as

$$\text{NMSE}(r,\hat{r}) = \frac{\sum_{n=0}^{N-1}(r[n]-\hat{r}[n])^2}{\sum_{n=0}^{N-1}(r[n]-\langle r\rangle)^2}, \tag{5.7}$$

where $\langle r \rangle$ represents the average reference firing rate. We note that this error measure violates the metric properties in Eq. (5.2) and in Eq. (5.3), due to the biased denominator term in Eq. (5.6).

5.3.3 Percent Variance Accounted For

Another error measure with an expression just slightly different from the NMSE is the *percent-Variance-Accounted-For*. The percent-Variance-Accounted-For (%VAF) is used to evaluate how well a model describes a biological system [Westwick and Kearney (2003); Pillow et al. (2005)]. In terms of firing rate, the %VAF compares the observed firing rate $r(t)$ with the estimated firing rate $\hat{r}(t)$ through the expression

$$\%\text{VAF}(r,\hat{r}) = 100 \times \frac{E\{[(r(t)-\hat{r}(t))-(\langle r\rangle-\langle \hat{r}\rangle)]^2\}}{E\{[r(t)-\langle r\rangle]^2\}}, \tag{5.8}$$

where the quantities between angle brackets are the mean firing rates of the observed and predicted trains that are calculated using Eq. (3.37). Expanding the expectation operator in Eq. (5.8), the %VAF becomes

$$\%\text{VAF}(r,\hat{r}) =$$
$$100 \times \frac{E\{r^2(t)\}+E\{\hat{r}^2(t)\}-\langle r\rangle^2-\langle \hat{r}\rangle^2+2\langle r\rangle\langle \hat{r}\rangle-2E\{r(t)\hat{r}(t)\}}{E\{r^2(t)\}-\langle r\rangle^2}, \tag{5.9}$$

where the last term in the numerator of Eq. (5.9) is the correlation between the observed and the estimated firing rates. If the two processes are independent, then $E\{r(t)\hat{r}(t)\} = E\{r(t)\}E\{\hat{r}(t)\} = \langle r\rangle\langle \hat{r}\rangle$, and the last two terms in Eq. (5.9) cancel each other. In this case, the two firing rates are uncorrelated, and the %VAF reduces to the sum of the variances of the observed and estimated firing rates normalized by the variance of the reference firing rate.

For a firing rate discretized into N time bins, this reduces to the expression

$$\%\text{VAF}(r,\hat{r}) = 100 \times$$
$$\frac{\frac{1}{N}\sum_{n=0}^{N-1} r^2[n] + \frac{1}{N}\sum_{n=0}^{N-1} \hat{r}^2[n] - \langle r \rangle^2 - \langle \hat{r} \rangle^2 + 2\left(\langle r \rangle \langle \hat{r} \rangle - \frac{1}{N}\sum_{n=0}^{N-1} r[n]\hat{r}[n]\right)}{\frac{1}{N}\sum_{n=0}^{N-1} r^2[n] - \langle r \rangle^2}.$$
(5.10)

Using the variance operator $(\text{VAR}(\cdot)^1)$, the expression for the %VAF metric can simply be written as

$$\%\text{VAF} = 100 \times \frac{\text{VAR}(r[n] - \hat{r}[n])}{\text{VAR}(r[n])}. \tag{5.11}$$

The numerator in Eq. (5.8) is the variance of the difference between the observed and the predicted firing rates, while that is not the case for the NMSE. Once again, the biased term in the denominator of Eq. (5.8) leads this error measure to violate Eq. (5.2) and Eq. (5.3).

5.3.4 *Analysis of the Firing Rate Metrics*

Table 5.1 summarizes the values obtained by applying the previously presented metrics to the salamander ON-type RGC responses. These values were obtained by dividing and comparing the set of 12 trials shown in Fig. 3.6(b) into two distinct sets, with each set composed of 6 trials. A large number of combinations of trials is possible, making it almost impossible to use all combinations in practice. The Monte-Carlo method [Flannery et al. (2002)] can be used to repeatedly randomly select 6 trials among the 12 trials. We can achieve this by indexing each trial, and randomly choosing 6 trials to form a set by picking the trials whose indices are obtained by sampling a uniform probability distribution.

Table 5.1 presents the errors obtained by applying the firing rate metrics to 1500 different combinations of randomly selected trials. From these results, it is possible to observe noticeable differences between the values of the MSE and its normalized version, the NMSE, and also between the use of straight PSTH and the smoothed PSTH.

If we are searching for a model that performs well in terms of the firing rate metrics, then the error values obtained by applying the metrics to its output should be kept within the range given by the values in Table 5.1.

---
[1] $\text{VAR}(x) = E\{(x-\mu_x)^2\} = E\{x^2\} - \mu_x^2$, where $\mu_x = E\{x\}$.

Table 5.1 Values of the firing rate metrics applied to salamander ON-type RGC responses.

Firing Rate Metrics			
Normal PSTH			
Metric	mean ± std	min	max
MSE	2454.6 ± 138.2447	2041.7	2875.0
NMSE	1.1450 ± 0.0542	0.9833	1.3234
%VAF	114.4976 ± 5.4213	98.3281	132.3359
Smoothed PSTH			
Metric	mean ± std	min	max
MSE	8.0686 ± 1.1295	5.5717	12.1673
NMSE	0.0593 ± 0.0088	0.0405	0.0951
%VAF	5.9054 ± 0.8836	4.0269	9.5090

A characteristic of the firing rate metrics is their sensitivity to phase (time shift). That is, by comparing two sets of spike trains whose only difference is a shift in the time occurrence of the peaks in their firing rates the error will be significant. This error could be even lower if the metric is applied to compare a firing rate obtained from a set of spike trains and a firing rate equal to zero, which is not a desirable result. Figure 5.2 displays the firing rates for two different sets of trials from the twelve trains, and it can be observed from Fig. 5.2(a) that the non smoothed PSTH possess big amplitude peaks so that a shift in the phase would result in a large increase in the error value. For the smoothed PSTHs of Fig. 5.2(b) the peaks' amplitudes are smaller and spread, so a phase shift only leads to a more moderate increase in the error (we must recall that the integral of the straight and of the smoothed PSTHs are equal, so that the difference for the metrics values are due to phase errors only). Therefore, the values in Table 5.1 obtained with the smoothed PSTHs are much lower precisely because smoothing decreases the errors due to phase differences.

5.4 Spike Train Metrics

Spike train metrics measure directly the (dis)similarity between two spike trains. These metrics are particularly suited for the assessment of time-code models by measuring the distance, or (dis)similarity, between the spike trains from a real retina and the ones predicted by a model. Three suitable metrics that can be found in the literature are: the *spike time metric*, the *interspike interval (ISI) metric*, and the *spike distance metric*. These metrics are employed to compare two spike trains, represented by

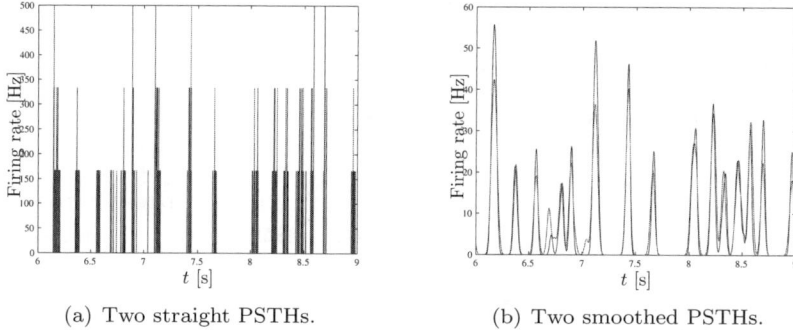

(a) Two straight PSTHs. (b) Two smoothed PSTHs.

Fig. 5.2 Comparing smoothed and not smoothed PSTHs.

their neural functions $\rho_1(t)$ and $\rho_2(t)$, composed of the spikes at time instants $t_{1,1}, t_{1,2}, \ldots, t_{1,n_1}$ and $t_{2,1}, t_{2,2}, \ldots, t_{2,n_2}$, respectively, as depicted in Fig. 5.3.

5.4.1 Spike Time Metric

A pair of spike train metrics was proposed in [Victor and Purpura (1996, 1997)], and revisited in [Victor (2005)]. These metrics consider the spike trains as points in a vector space, and the metric is applied to calculate their distance. The *spike time metric*, denoted by $d^{\text{time}}(\cdot)$, considers the absolute time occurrence of the individual spikes in a train as the significant quantity that distinguishes two spike trains.

The distance between spike trains $\rho_1(t)$ and $\rho_2(t)$ is measured as the minimum cost path followed to transform one spike train into the other by iteratively applying a set of elementary operations. The cost to transform a spike train $\rho_i(t)$ into $\rho_j(t)$, like the ones depicted in Fig. 5.3 (where $i = 1$

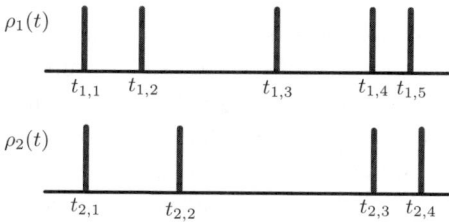

Fig. 5.3 Comparison of two spike trains.

and $j = 2$), denoted by the distance $d(\rho_i(t), \rho_j(t))$, is equal to the sum of the individual costs associated with a particular sequence of elementary steps that successively transform the spike train $\rho_i(t)$ into $\rho_j(t)$:

$$d(\rho_i(t), \rho_j(t)) = \min \{d(\rho_i(t), \rho_{i+1}(t)) + \cdots + d(\rho_{j-1}(t), \rho_j(t))\}, \quad (5.12)$$

where the difference between $\rho_i(t)$ and $\rho_{i+1}(t)$ corresponds to a single elementary operation. The minimum is calculated over all possible paths of spike trains, $\rho_i(t), \rho_{i+1}(t), \ldots, \rho_{j-1}(t), \rho_j(t)$, beginning in $\rho_i(t)$ and ending at $\rho_j(t)$.

The metric $d^{\text{time}}(\cdot)$ is a function of a parameter q that measures its sensitivity to the timing of the spikes occurrence. The permitted elementary operations are: i) the insertion and ii) the deletion of an individual spike, both of these operations have a unitary cost per spike; and iii) the shift in the time occurrence of a spike, with a cost of q per second, which leads to a cost of $q|\Delta t|$ for a shift of Δt seconds on the occurrence of a spike. The further away a spike is relative to the spike on the other train, the more costly it is to shift the spike to the right place. The shifting of a spike can be so costly that it can be less costly to delete it and raise another spike coincident with the comparing spike, which has a total cost of 2 (d^{time}(spike delete + spike raise) = 2).

Extending the previous reasoning, we can find two antipodal limit cases. The first case occurs when $q = 0$, meaning that the shifting of a spike is costless, so that the contributions to the cost are due only to the deletion or raising of spikes. If the spike train $\rho_1(t)$, with n_1 spikes, is being compared with the spike train $\rho_2(t)$, with n_2 spikes, the minimum path to transform one spike train into the other can be done with $|n_1 - n_2|$ steps by creating (or deleting) a spike at each step with a cost of $|n_1 - n_2|$. This corresponds to a spike-count metric, $d^{\text{time}}_{q=0}(\cdot) = d^{\text{count}}(\cdot)$, where the only relevant characteristic of the spike trains is its total number of spikes. According to d^{count}, two spike trains with the same number of spikes have an error equal to zero.

The other extreme case occurs when the value of q is very high. For example, consider two spike trains with a single spike each positioned at times t and t', respectively, the cost for shifting the spike to the right place is $d^{\text{time}}(\rho_1(t), \rho_2(t)) = q|t - t'|$. The alternative path is to delete the spike at the time instant t and create another spike at time t', which has a total cost of 2. The latter path can be less costly than the former if $|t - t'| > 2/q$. When q is very high, ($q \to \infty$) the distance between the two spike trains $\rho_1(t)$ and $\rho_2(t)$ is $d^{\text{time}}_{q \to \infty} = n_1 + n_2 - 2c$, where c is the number of coincident

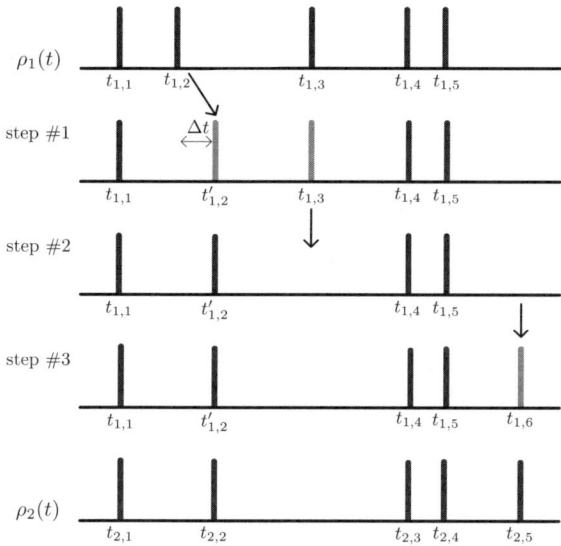

Fig. 5.4 Path to transform a spike train into the other in the spike time metric.

spikes in time on the two trains under comparison.

In between the two extreme situations described, $d_q^{\text{time}}(\cdot)$ defines a family of metrics where displacing a spike by an amount of $1/q$ is equal to the cost of deleting it; q is a measure of the precision of the temporal coding and establishes how far a spike can be shifted without increasing the distance between the two spike trains with regard to $d_{q\to\infty}^{\text{time}}(\cdot)$.

In the search for a minimal path between two spike trains, there are restrictions in the sequence of operations that can be used. These restrictions include: the path cannot comprise the shift and deletion of the same spike, because the cost will be lower by just deleting the spike; a spike cannot be included and then shifted, since the direct insertion of the spike in its right position is less costly; and the insertion and deletion of the same spike cannot be part of the minimal path between two spike sequences. Also, a minimal path cannot include both leftward and rightward spike shifts of a spike, nor the crossing of two spikes. This reduces the number of operations that can be exerted on a certain spike when comparing trains $\rho_1(t)$ and $\rho_2(t)$ to three: *i)* deletion of the last analyzed spike in $\rho_1(t)$, *ii)* insertion of a spike in $\rho_1(t)$, or *iii)* the two last spikes in $\rho_1(t)$ and $\rho_2(t)$ are related by a shift. Figure 5.4 illustrates a set of possible transforming operations that can be exerted to transform one spike train into the other, where in the first step a spike is shifted by Δt in train $\rho_1(t)$, in the second step a spike is deleted from $\rho_1(t)$, and in the third step a spike is inserted in the train, so that we arrive at $\rho_2(t)$.

Example 5.1. Calculate the distance between the two spike trains $\rho_1(t)$ and $\rho_2(t)$ from Fig. 5.4, with the cost of a spike shift equal to zero ($q = 0$).

Since the cost to shift a spike is null we can match the spikes from $\rho_1(t)$ with the ones from $\rho_2(t)$ by shifting them without any cost. If we follow this procedure we can match all spikes in time in the two trains with a total cost of $d_{q=0}^{\text{time}}(\rho_1, \rho_2) = 0$. Because the two trains have the same number of spikes, neither deletion nor insertion of spike must be performed. In this particular case, we could have directly applied the formula $d_{q=0}^{\text{time}}(\rho_1, \rho_2) = |5-5| = 0$.■

Example 5.2. By considering a very high cost for the spike shift ($q \to \infty$), obtain the distance between the spike trains $\rho_1(t)$ and $\rho_2(t)$ from Fig. 5.4 according to the spike time metric.

With a very high value for q in the spike time metric, a spike is never shifted, and therefore only deletion and insertion operations are performed. Since the occurrence times of the first spikes in $\rho_1(t)$ and $\rho_2(t)$ match, that is $t_{1,1} = t_{2,1}$, there is no cost to make these spikes coincide. Concerning the second spike, it is less costly to delete the spike at $t_{1,2}$ and raise another one in $t'_{1,2} = t_{2,2}$, with a total cost of 2. The third spike in $\rho_1(t)$ is just deleted with cost 1. Since the fourth and fifth spikes in $\rho_1(t)$ coincide with the third and fourth spikes in $\rho_2(t)$, respectively, there is no cost to make them match. Finally, we have to raise a spike at time $t_{1,6}$ to match the one at $t_{2,5}$ in $\rho_2(t)$, which has cost equal to 1. Summing all these individual costs, we have that $d_{q \to \infty}^{\text{time}}(\rho_1, \rho_2) = 4$. We could also have employed the formula $d_{q \to \infty}^{\text{time}}(\rho_1, \rho_2) = n_1 + n_2 - 2c$, where $n_1 = 5$ and $n_2 = 5$, and the number of coincident spikes is $c = 3$.■

The process to calculate the distance $d^{\text{time}}(\rho_1(t), \rho_2(t))$ between two spike trains can be formalized in Algorithm 5.1. Let's consider that the spike train $\rho_1(t)$ has n_1 spikes located at times $\{t_{1,1}, t_{1,2}, \ldots, t_{1,n_1}\}$, and $\rho_2(t)$ has n_2 spikes located at $\{t_{2,1}, t_{2,2}, \ldots, t_{2,n_2}\}$ (see Fig. 5.4). If $d_{i,j}^{\text{time}}$ is the spike time distance between the two spike trains composed of the first i spikes of train $\rho_1(t)$ and of the first j spikes of train $\rho_2(t)$, the allowed

Algorithm 5.1 Spike Time Metric, d^{time}

1: $\mathbf{t}_1 \leftarrow$ vector of spike time instants for train ρ_1
2: $\mathbf{t}_2 \leftarrow$ vector of spike time instants for train ρ_2
3: $n_1 \leftarrow$ number of elements in $\mathbf{t}_1$, i.e., number of spikes in train ρ_1
4: $n_2 \leftarrow$ number of elements in $\mathbf{t}_2$, i.e., number of spikes in train ρ_2
5: $q \leftarrow$ penalty to shift one spike by one second (in any direction)
6: $\mathbf{D} \leftarrow$ matrix of size $[(n_1 + 1) \times (n_2 + 1)]$ where
   - first row: $\mathbf{D}_{0,:} = [0, \cdots, n_2]$
   - first column: $\mathbf{D}_{:,0} = [0, \cdots, n_1]^T$
7: **for** $i = 1$ to n_1 **do** {*For all spikes in train ρ_1, $\mathbf{t}_1$*}
8:   **for** $j = 1$ to n_2 **do** {*For all spikes in train ρ_2, $\mathbf{t}_2$*}
9:     $\mathbf{D}_{i,j} = \min \left\{ \mathbf{D}_{i-1,j} + 1 \, , \, \mathbf{D}_{i,j-1} + 1, \mathbf{D}_{i-1,j-1} + q|\mathbf{t}_{1,i} - \mathbf{t}_{2,j}| \right\}$
10: **end for**
11: **end for**
12: $d^{\text{time}} \leftarrow \mathbf{D}_{n_1,n_2}$
---

operations listed above imply that the minimum path is

$$d^{\text{time}}_{i,j} = \min \left\{ \underbrace{d^{\text{time}}_{i-1,j} + 1}_{\substack{\text{Add spike to } \rho_1 \text{ at } t_{2,j} \\ \text{(Erase spike in } \rho_2 \text{ at } t_{2,j})}} \; ; \; \underbrace{d^{\text{time}}_{i,j-1} + 1}_{\substack{\text{Add spike to } \rho_2 \text{ at } t_{1,i} \\ \text{(Erase spike in } \rho_1 \text{ at } t_{1,i})}} \; ; \; \underbrace{d^{\text{time}}_{i-1,j-1} + q|t_{1,i} - t_{2,j}|}_{\substack{\text{Shift spike in } \rho_1 \text{ from } t_{1,i} \text{ to } t_{2,j} \\ \text{(Shift spike in } \rho_2 \text{ from } t_{2,j} \text{ to } t_{1,i})}} \right\}. \qquad (5.13)$$

By successively applying the expression in Eq. (5.13) to every new spike in each spike train $\rho_1(t)$ and $\rho_2(t)$, we can construct a two-dimensional array, where the cell in the ith row and jth column contains the distance $d^{\text{time}}_{i,j}$. The first row of this two-dimensional array is filled with $d^{\text{time}}_{0,j} = j$, corresponding to have zero spikes in the first train and j spikes in the second train, so that all spikes are created in the first train (or deleted in the second train). The first column of this array is filled with $d^{\text{time}}_{i,0} = i$, corresponding to i spikes in ρ_1 and zero spikes in ρ_2, so that all spikes are deleted on the first train (or created in the second train). By iteratively filling the $(n_1 + 1) \times (n_2 + 1)$ array, the overall cost $d^{\text{time}}(\rho_1(t), \rho_2(t))$ is located in the array cell (n_1, n_2). The Algorithm 5.1 summarizes the steps needed to implement this metric in a computer, where the matrix $\mathbf{D}_{(n_1+1,n_2+1)}$ is recursively filled with the distances given by Eq. (5.13). This procedure was first proposed and established to compare gene sequences [Sellers (1974)], and is also used to measure distances between strings. This is known as the edit, or Levenshtein, distance.

Example 5.3. Calculate the distance between the two spike trains $\rho_1(t)$ and $\rho_2(t)$ from Fig. 5.4 using the spike time metric for a cost per second of spike shift equal to q.

The overall error is equal to the sum of the individual costs for each basic operation performed in each step. Therefore, in step #1, the cost to shift the spike from $t_{1,2}$ to $t'_{1,2} = t_{1,2} + \Delta t$ is $q\Delta t$. In step #2, a spike is deleted, which has cost equal to 1, and in step #3, a spike is inserted, which has cost 1. Thus, the overall cost is $d_q^{\text{time}}(\rho_1, \rho_2) = q\Delta t + 2$. ■

5.4.2 Interspike Interval Metric

The time intervals between two consecutive spikes convey the relevant information in the *interspike interval metric*, $d^{\text{inter}}(\cdot)$, for trains comparison. The neurobiological justification to use this metric is related to the fact that, due to the dynamics of the ions' channels in the neuron, the effect of an action potential depends on the length of the time interval since the last spike was fired [Victor and Purpura (1996, 1997)].

Similar to the spike time metric, the allowed elementary operations are the addition of an interspike interval with the associated spike and the removal of an interspike interval with the associated spike; both operations have a cost equal to 1. The other allowed operation is the shift of a spike, that changes the duration of the associated ISI, with a cost q per second. As illustrated in Fig. 5.5 a set of elementary steps are applied to transform the spike train $\rho_1(t)$ into $\rho_2(t)$. The deletion and insertion of an interspike interval are exemplified in the second and third steps of Fig. 5.5, respectively, while the shift of a spike, with the consequent change of its interval duration and the shift of all subsequent spikes, is exemplified in the first and fourth steps. Note that changing an ISI duration by the amount Δt has a cost of $q|\Delta t|$, and has the consequence of changing the time occurrence of all subsequent spikes by Δt; however, their interspike intervals remain unchanged.

Like in the spike time metric, for $q = 0$, this metric degenerates to the spike count metric: $d_{q=0}^{\text{inter}}(\cdot) = d^{count}(\cdot)$. Since the cost of changing the duration of an ISI is null ($q = 0$), the distance is due only to the removal

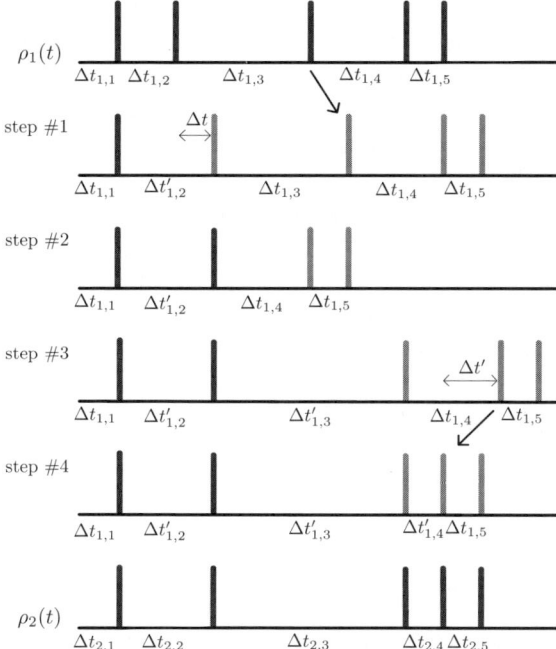

Fig. 5.5 Path to transform a spike train into another in the interspike interval metric.

(or insertion) of ISIs with the associated spikes, so that $d^{\text{inter}}_{q=0}(\rho_1, \rho_2) = d^{count}(\rho_1, \rho_2) = |n_1 - n_2|$, for two trains ρ_1 and ρ_2 with n_1 and n_2 spikes, respectively.

Example 5.4. Employing the interspike interval metric, calculate the distance between the spike trains $\rho_1(t)$ and $\rho_2(t)$ from Fig. 5.4 for a cost of interval duration adjustment equal to zero ($q = 0$).

Since the cost to change the time length of an interval is null, the interspike intervals of $\rho_1(t)$ and $\rho_2(t)$ can be adjusted without any cost. Thus, since the number of ISIs are equal in both trains, adjusting all interval durations we are left without any interval and so $d^{\text{inter}}_{q=0}(\rho_1, \rho_2) = |5 - 5| = 0.$ ■

For values of q different from zero, the distance depends on the temporal pattern of spikes composing the sequence, namely on their ISIs. For $q \to \infty$ an ISI duration will never be adjusted since it is less costly to remove the spike with its ISI, and insert another ISI with the right duration, which has a total cost of 2 (d^{inter}(ISI removal + ISI insertion) = 2), and only the coincident ISIs between the two trains will not be replaced. Therefore, similar to the spike time metric, the distance between the spike trains ρ_1 and ρ_2, with n_1 spikes and n_2 spikes (and consequently with n_1 and n_2 spike intervals, respectively), for $q \to \infty$ will be $d_{q\to\infty}^{inter}(\rho_1, \rho_2) = n_1 + n_2 - 2c$, where c is the number of interspike intervals with the same duration. Two ISIs are equal if the interspike interval before spike i in train ρ_1 is equal to the interspike interval before spike j in ρ_2: $\Delta t_{1,i} = \Delta t_{2,j}$. For two consecutive interspike intervals that are equal by $\Delta t_{1,i} = \Delta t_{2,j}$ and $\Delta t_{1,k} = \Delta t_{2,l}$, we must have $k > i$ and $l > j$, because we cannot reverse the order of the intervals. It could happen that we have a different number of coincident ISIs depending on the matching sequence.

Example 5.5. By employing the interspike interval metric, calculate the distance between the spike trains $\rho_1(t)$ and $\rho_2(t)$ from Fig. 5.4 when the cost of changing an interspike time interval is very high ($q \to \infty$).

In the interspike interval metric, a very high value for q means that only removal and insertion of ISIs operations are applied. Hence, we have to transform the train $\rho_1(t)$ into $\rho_2(t)$ solely with these operations. By assuming that the first interspike intervals are equal ($\Delta t_{1,1} = \Delta t_{2,1}$) there is no cost to transform one into the other. Comparing the second interspike intervals, we have to delete the ISI $\Delta t_{1,2}$ (cost 1), and insert another ISI with duration $\Delta t'_{1,2} = \Delta t_{2,2}$ (cost 1). Then we have to match $\Delta t_{1,3}$ with $\Delta t_{2,3}$, but since they are different, the only way is to remove $\Delta t_{1,3}$ (cost 1), and insert another ISI with duration $\Delta t'_{1,3} = \Delta t_{2,3}$ (cost 1). The interspike intervals $\Delta t_{1,4}$ and $\Delta t_{2,4}$ are also different, and we have to remove $\Delta t_{1,4}$ (cost 1), and insert another ISI with duration $\Delta t'_{1,4} = \Delta t_{2,4}$ (cost 1). Finally, the intervals $\Delta t_{1,5}$ and $\Delta t_{2,5}$ are equal and there is no cost to transform one into the other. The overall distance, which can be obtained by summing the individual costs of the previous described operations, is $d_{q\to\infty}^{inter}(\rho_1, \rho_2) = n_1 + n_2 - 2c = 5 + 5 - 2 \times 2 = 6$, because there are two coincident intervals: $\Delta t_{1,1} = \Delta t_{2,1}$ and $\Delta t_{1,5} = \Delta t_{2,5}$. Note that $d_{q\to\infty}^{inter}(\rho_1, \rho_2) = 6$ gives the total number of interspike intervals that have to be removed, or inserted. ■

Similar to the case of the spike time metric, we can derive Algorithm 5.2 to iteratively calculate the distance between two spike trains based on the time intervals between successive spikes [Victor and Purpura (1996); Sellers (1974)]. Considering the spikes in $\rho_1(t)$ located at times $\{t_{1,1}, t_{1,2}, \ldots, t_{1,n_1}\}$, and in $\rho_2(t)$ located at $\{t_{2,1}, t_{2,2}, \ldots, t_{2,n_2}\}$, spike intervals are defined for ρ_1 by $\Delta t_{1,i} = t_{1,i} - t_{1,i-1}$, and for ρ_2 by $\Delta t_{2,j} = t_{2,j} - t_{2,j-1}$ (see Fig. 5.5). Denoting by $d_{i,j}^{\text{inter}}$ the interspike interval distance between two spike trains composed of the first i spikes of train $\rho_1(t)$ and the first j spikes of the train $\rho_2(t)$, the allowed operations imply that the minimum path can be constructed iteratively by applying the expression:

$$d_{i,j}^{\text{inter}} = \min \left\{ \underbrace{d_{i-1,j}^{\text{inter}} + 1}_{\substack{\text{Add ISI } \Delta t_{2,j} \text{ to } \rho_1 \\ (\text{Remove ISI } \Delta t_{2,j} \text{ in } \rho_2)}} \; ; \; \underbrace{d_{i,j-1}^{\text{inter}} + 1}_{\substack{\text{Add ISI } \Delta t_{1,i} \text{ to } \rho_2 \\ (\text{Remove ISI } \Delta t_{1,i} \text{ in } \rho_1)}} \; ; \; \underbrace{d_{ij}^{\text{inter}} + q|\Delta t_{1,i} - \Delta t_{2,j}|}_{\substack{\text{Change ISI } \Delta t_{1,i} \text{ in } \rho_1 \text{ to } \Delta t_{2,j} \\ (\text{Change ISI } \Delta t_{2,j} \text{ in } \rho_2 \text{ to } \Delta t_{1,i})}} \right\}.$$
(5.14)

By successively applying this procedure, a two-dimensional array, with dimensions $(n_1 + 1) \times (n_2 + 1)$, can be constructed with the interspike distances $d_{i,j}^{\text{inter}}$ between the first i ISIs of $\rho_1(t)$ and the first j ISIs of $\rho_2(t)$. Algorithm 5.2 summarizes the steps needed to implement this metric in a computer, where the matrix $\mathbf{D}_{(n_1+1,n_2+1)}$ is iteratively filled with the distances given by Eq. (5.14), and the overall distance $d_{\rho_1,\rho_2}^{\text{inter}}$ ends in the element (n_1, n_2) of the matrix $\mathbf{D}$. The duration of the first ISIs are unknown, since we do not known when the previous spikes occurred. However, they must be at least equal to the time between the start of the data recording until the first spike occurs, so we can make $\Delta t_{1,1} = t_{1,1}$ and $\Delta t_{2,1} = t_{2,1}$ to initialize the algorithm; there are other possibilities [Victor and Purpura (1997)]. The first row of the two dimensional array can be filled with $d_{0,j}^{\text{inter}} = j$, meaning that all interspike intervals have to be removed from ρ_1 (or inserted in ρ_2), while the first column can be filled by taking into account that $d_{i,0}^{\text{inter}} = i$, meaning that all interspike intervals have to be removed from ρ_2 (or inserted in ρ_1).

Example 5.6. Obtain the distance between the spike trains $\rho_1(t)$ and $\rho_2(t)$ from Fig. 5.5 using the interspike interval metric by considering the cost per second for changing an interspike duration to be q.

The overall error is equal to the sum of the individual costs for each basic operation performed in each step. Therefore, in step #1, the interspike $\Delta t_{1,2}$ is changed to $\Delta t'_{1,2} = \Delta t_{1,2} + \Delta t$ to match the interval $\Delta t_{2,2}$, with a cost of $q\Delta t$. In step #2, the interval $\Delta t_{1,3}$ is removed with the associated spike, with a cost 1, and in step #3, the interspike interval $\Delta t'_{1,3}$ is inserted,

Algorithm 5.2 Interspike interval metric, d^{inter}

1: $\mathbf{t}_1 \leftarrow$ vector of spike time instants for train ρ_1
2: $\mathbf{t}_2 \leftarrow$ vector of spike time instants for train ρ_2
3: $n_1 \leftarrow$ number of elements in $\mathbf{t}_1$, i.e., number of spikes in train ρ_1
4: $n_2 \leftarrow$ number of elements in $\mathbf{t}_2$, i.e. number of spikes in train ρ_2
5: $q \leftarrow$ penalty to increase/decrease one second the interspike interval
6: $\mathbf{D} \leftarrow$ matrix of size $[(n_1+1) \times (n_2+1)]$ where
   - first row: $\mathbf{D}_{0,:} = [0, \cdots, n_2]$
   - first column: $\mathbf{D}_{:,0} = [0, \cdots, n_1]^T$
7: $\{\mathbf{I}_1, \mathbf{I}_2\} \leftarrow$ vector of interspike intervals for trains ρ_1 and ρ_2, respectively:
   - $\mathbf{I}_{1,1} \leftarrow \mathbf{t}_{1,1}$; $\mathbf{I}_{2,1} \leftarrow \mathbf{t}_{2,1}$;
   - $\mathbf{I}_{1,i} \leftarrow \mathbf{t}_{1,i} - \mathbf{t}_{1,i-1}$, $\mathbf{I}_{2,j} \leftarrow \mathbf{t}_{2,j} - \mathbf{t}_{2,j-1}$, $\forall_{i,j>1}$
8: **for** $i = 1$ to n_1 **do** {For all spikes in train ρ_1, $\mathbf{t}_1$}
9:     **for** $j = 1$ to n_2 **do** {For all spikes in train ρ_2, $\mathbf{t}_2$}
10:        $\mathbf{D}_{i,j} = \min\left\{\mathbf{D}_{i-1,j}+1\ ,\ \mathbf{D}_{i,j-1}+1, \mathbf{D}_{i-1,j-1}+q|\mathbf{I}_{1,i}-\mathbf{I}_{2,j}|\right\}$
11:    **end for**
12: **end for**
13: $d^{\text{inter}} \leftarrow \mathbf{D}_{n_1,n_2}$

which has a cost of 1. In step #4, the duration of the interspike interval $\Delta t_{1,4}$ is changed to $\Delta t'_{1,4} = \Delta t_{1,4} - \Delta t'$ which has a cost $q\Delta t'$. Summing all the individual costs, the distance between $\rho_1(t)$ and $\rho_2(t)$ following the particular path shown in Fig. 5.5 is $d_q^{\text{inter}}(\rho_1, \rho_2) = q(\Delta t + \Delta t') + 2$. ∎

5.4.3 Spike Train Distance Metric

The *spike train distance metric*, $d^2(\cdot)$ is a spike train metric less general than the previous ones, but easier to analyze. This metric measures the dissimilarity between two spike trains by taking into account the temporal structure of the train, and depends on a temporal parameter τ that measures the contribution of the displacement of spikes in the two trains under comparison. Authors in [van Rossum (2001)] claim that this metric is not unlikely to be implemented physiologically and is simpler than the previous ones. The spike train distance metric compares a modified version of the spike trains resulting from the convolution of the neural function with a exponentially decaying kernel. This kernel can be interpreted as the post-synaptic potentials, in higher order-neurons for a small decay rate, while for a large decaying rate it can be interpreted as calcium-induced currents in the neuron membrane [Dayan and Abbot (2001)].

The metric modifies the neural response function $\rho(t)$, given by Eq. (3.7), by replacing every spike represented as a delta functions at times t_i, $i = 1, \ldots, n$, by a decaying exponential function:

$$h(t) = \mathrm{u}(t) \exp^{-t/\tau}, \tag{5.15}$$

where the decay rate is τ. The modified spike train becomes

$$\rho'(t) = \rho(t) * h(t) = \sum_{i=1}^{n} \mathrm{u}(t - t_i) \, e^{-(t-t_i)/\tau}. \tag{5.16}$$

Figure 5.6 displays the schematic result of the convolution of spike trains with the exponentially decaying kernel. According to the spike distance metric, the distance between the two spike trains, $\rho_1(t)$ and $\rho_2(t)$, is defined by the expression

$$d^2(\rho_1(t), \rho_2(t)) = \frac{1}{\tau} \int_0^\infty [\rho'_1(t) - \rho'_2(t)]^2 \, dt, \tag{5.17}$$

which corresponds to the Euclidean distance between the two filtered spike trains $\rho'_1(t)$ and $\rho'_2(t)$. In this expression, the modified spike trains $\rho'_1(t)$ and $\rho'_2(t)$ are subtracted, and the square of those values are integrated in time.

The metric shows a different behavior depending on the values for the decay constant τ. For a value of τ much smaller than the ISI, the smoothed spike trains in Eq. (5.16) will contribute to the integral only if the spikes in both sequences are separated by more than τ, and the metric works as a coincidence detector. In the case of continuous time spike trains, the number of coincident spikes can be neglected when $\tau \to 0$, so that the cross terms obtained by expanding the integrand in Eq. (5.17) can be neglected, and the expression for the distance becomes

$$\lim_{\tau \to 0} d^2(\rho_1(t), \rho_2(t)) = \frac{1}{\tau} \int_0^\infty [\rho'^2_1(t) + \rho'^2_2(t)] \, dt = \frac{1}{2}(n_1 + n_2), \tag{5.18}$$

where n_1 and n_2 are the number of spikes in the trains $\rho_1(t)$ and $\rho_2(t)$, respectively. Thus, the metric just counts the number of noncoincident spikes. If the spike trains have c coincident spikes, the distance metric gives

$$\lim_{\tau \to 0} d^2(\rho_1(t), \rho_2(t)) = \frac{1}{2}(n_1 + n_2 - 2c), \tag{5.19}$$

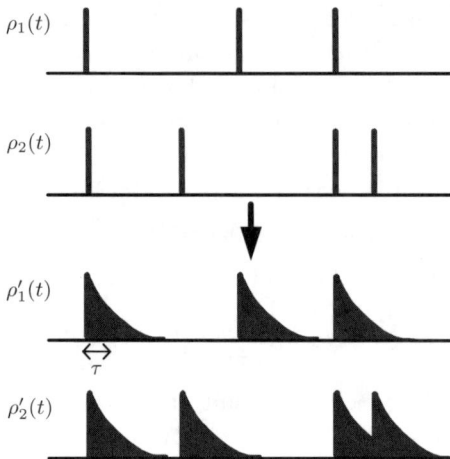

Fig. 5.6 Changing of the spike shape for comparison with the spike distance metric.

which is equal (ignoring the multiplicative factor 1/2) to the result obtained with the spike time metric when a large cost for the spike shifting operation ($d_{q \to \infty}^{\text{time}}$) is considered.

On the other hand, when we use a very large value for τ ($\tau \to \infty$), the main contribution to the integral of Eq. (5.17) comes from the times when the last spike has passed but the exponent has still not decayed. In that case, the metric can be approximated by

$$\lim_{\tau \to \infty} d^2(\rho_1(t), \rho_2(t)) = \lim_{\tau \to \infty} \frac{1}{\tau} \int_0^\infty \left(\sum_{i=1}^{n_1} u(t - t_i) e^{-(t-t_i)/\tau} \right.$$
$$\left. - \sum_{j=1}^{n_2} u(t - t_j) e^{-(t-t_j)/\tau} \right)^2 dt \qquad (5.20)$$
$$= \lim_{\tau \to \infty} \frac{1}{\tau} \int_0^\infty \left(n_1 e^{-t/\tau} - n_2 e^{-t/\tau} \right)^2 dt$$
$$= \frac{1}{2} (n_1 - n_2)^2.$$

The second equality results from the fact that

$$\lim_{\tau \to \infty} \sum_{i=1}^n u(t - t_i) e^{-(t-t_i)/\tau} = \sum_{i=1}^n \underbrace{\lim_{\tau \to \infty} u(t - t_i) e^{t_i/\tau}}_{=1} e^{-t/\tau} \qquad (5.21)$$

Expressions for some special operations on the spike trains, like the insertion (or deletion) of a spike, and the shift of a spike can also be derived. If the spike train $\rho_2(t)$ differs from $\rho_1(t)$ just for a spike placed at time t_i, such that their convolved versions are related by

$$\rho'_2(t) = \rho'_1(t) + \mathrm{u}(t - t_i)e^{-(t-t_i)/\tau}, \qquad (5.22)$$

the distance metric gives

$$d^2\bigl(\rho_1(t), \rho_2(t)\bigr) = \frac{1}{\tau} \int_{t_i}^{\infty} e^{-2(t-t_i)/\tau} dt = \frac{1}{2}, \qquad (5.23)$$

and the deletion, or removal, of a spike produces the same value that is independent of τ.

For two spike trains whose only difference is the shift of a spike from t_i to $t_i + \Delta t$ in train $\rho_2(t)$ relative to $\rho_1(t)$, such that the convolved spike trains are related by

$$\rho'_2(t) = \rho'_1(t) - \mathrm{u}(t - t_i)e^{-(t-t_i)/\tau} + \mathrm{u}(t - t_i - \Delta t)e^{-(t-t_i-\Delta t)/\tau}, \qquad (5.24)$$

the distance between these spike trains is

$$\begin{aligned} d^2\bigl(\rho_1(t), \rho_2(t)\bigr) &= \frac{1}{\tau} \int_{t_i}^{t_i+\Delta t} e^{-2(t-t_i)/\tau} dt \\ &\quad + \frac{1}{\tau} \int_{t_i+\Delta t}^{\infty} \left[e^{-(t-t_i)/\tau} - e^{-(t-t_i-\Delta t)/\tau}\right]^2 dt \\ &= 1 - e^{-|\Delta t|/\tau}, \end{aligned} \qquad (5.25)$$

which approaches the value one for a large distance Δt between spikes compared to the decay τ, and zero when this distance is small compared to τ. Analytical expressions can be derived for other particular relations between the spike trains under analysis [van Rossum (2001); Tomás and Sousa (2008)]. An interesting case considers the distance between two uncorrelated homogeneous Poisson spike trains generated with the same constant firing rate r. According to Eq. (5.18) the distance between two spike trains approaches $(n_1 + n_2)/2$ for a small τ, where n_1 and n_2 are the number of spikes in each trial. However, as stated by Eq. (3.75), the number of expected spikes in a Poisson trial with a constant firing rate r and a time duration T is equal to rT, so that

$$d^2_{\tau \to 0}\bigl(\rho_1(t), \rho_2(t)\bigr) = rT. \qquad (5.26)$$

On the other hand, for large τ, the distance values tend towards $(n_1 - n_2)^2/2$, as stated by Eq. (5.20). The expectation value $\langle(n_1 - n_2)^2\rangle/2 = \langle n_1^2\rangle/2 + \langle n_2^2\rangle/2 - \langle n_1 n_2\rangle$ for a Poisson process can be calculated with the help of Eq. (3.72) and Eq. (3.73). Thus, $\langle n_1^2\rangle = \langle n_2^2\rangle = rT + (rT)^2$ and $\langle n_1 n_2\rangle = (rT)^2$, which results in

$$d_{\tau \to \infty}^2 (\rho_1(t), \rho_2(t)) = rT. \tag{5.27}$$

From Eq. (5.26) and Eq. (5.27) we observe that the average value for the distance metric is equal for small and for large values of τ in the case of a spike train described by a homogeneous Poisson process with mean firing rate r.

This metric can be computed by convolving the involved discrete spike trains, $\rho_1[n]$ and $\rho_2[n]$, with the discrete kernel

$$h[n] = \mathrm{u}[n]e^{-nT_s/\tau} \tag{5.28}$$

where $\mathrm{u}[n]$ is the discrete Heaviside unit step function and T_s is the sampling period. The convolved spike trains are subtracted from each other, and the result is squared and summed (the discrete equivalent to integration). The final expression for the discrete spike train distance is

$$d^2(\rho_1[n], \rho_2[n]) = \frac{1}{\tau}\sum_{i=0}^{N}(\rho_1'[i] - \rho_2'[i])^2 T_s, \tag{5.29}$$

where $\rho'[n] = \rho[n] * h[n]$ and N is the length of the discretized spike trains.

5.4.4 Spike Train Metrics Analysis

Originally, the spike time metric and the spike interval metric were applied to classify a set of neural responses, namely with the objective of finding the metric that aggregates the neural responses in a more compact subspace of the response space [Victor and Purpura (1996)]. The metrics gave different results and was noted by its proponents that the spike time metric and the interspike interval metric do not refine each other in the topological sense. This means that a sequence of successive steps that minimize the distance between two spike trains according to one metric can have the opposite effect with respect to the other metric [Victor and Purpura (1997)].

Figure 5.7 presents two graphics that show the evolution of the spike time metric and the interspike interval metric as a function of q by comparing the twelve trials in Fig. 3.6(b). Table 5.2 displays the values for the maxima and minima for the limit values of q. For $q = 0$ the minimum and

Table 5.2 Limit values for the spike train metrics using the neuronal responses of a ON-type salamander RGC.

	Spike Train Metrics			
	Spike Time Metric		Interspike Interval Metric	
q	min	max	min	max
$q = 0$	0	9	0	9
$q = \infty$	132	165	126	147

maximum values are equal for both metrics. In this case ($q = 0$), both metrics reduce to the spike count metric – $d_{q=0}^{\text{time}}(\rho_1, \rho_2) = d_{q=0}^{\text{inter}}(\rho_1, \rho_2) = d^{\text{count}}(\rho_1, \rho_2)$. For $q = 0$, both metrics show a zero value for the minimum because there are several spike trains within the set of analyzed trials with the same number of spikes. For the case of the maximum value, two spike trains can differ, in terms of the number of spikes, by 9 at most. For $q \to \infty$, the values for the spike time metric indicate that there are, at the minimum, 132 and at the maximum 165 non-coincident spikes in the spike trains. For the interspike interval metric, the values for $q \to \infty$ mean that the spike trains under comparison possess 126 noncoincident spike intervals at minimum, and 147 at maximum.

The curves in Fig. 5.7 should remain constant for values of $q \geq 2/T_s$, where T_s is the sampling period ($T_s = 1$ ms). This value corresponds to the situation where the cost to delete and raise a spike is lower (or the same for the case of the equal sign) than shifting it, even when the shift is only by one sampling period. As the graphics of Fig. 5.7 show, the distances between any two spike trains are constant for $q \geq 1$, meaning that no two trains in this experimental data have spikes that are displaced by only one

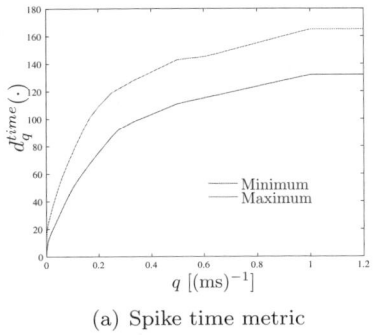

(a) Spike time metric

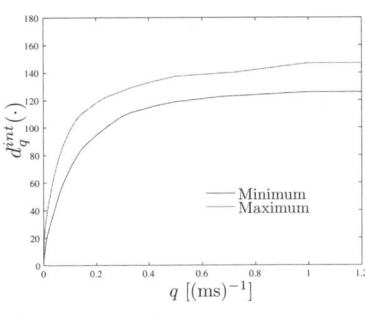

(b) Interspike interval metric

Fig. 5.7 Minimum and maximum evolution of the spike time and interspike metrics between a set of spike trains for a salamander ON-type RGC as a function of q.

time bin shift of T_s (they are at least displaced by two time bins).

The values of the metrics are different due to their distinct processes of deleting and inserting a spike (the only operations permitted for $q \to \infty$). For the spike time metric, a spike is just replaced by an empty bin, or a spike is placed in a certain time bin when raised. On the other hand, for the interspike interval metric, the removal (or insertion) of a spike has the consequence of shifting all subsequent spikes to the left (or to the right, respectively).

From Fig. 5.7 we can also have insight into the values for q and, for a given value of q, the range of values for the error between the observed cell's spike trains and those produced by a given model. The desired values for the error between the observed spike trains and the spike trains estimated from a model should be located between the curves in Fig. 5.7, for the appropriate metric, and for a given value of q.

An enlightening exercise is to choose a trial from the ON-type salamander cell and measure its distance to a spike train with no spikes at all – a null spike train. By repeating this exercise for each trial of the responses from the salamander ON-type cell, the minimum distance obtained is 78, what is equivalent to creating (or destroying) all spikes in the trial with fewer spikes, and the maximum distance obtained is 87, which is equivalent to destroying (or creating) all spikes in the train with more spikes. By comparing these values with the curves of Fig. 5.7, we can see that for values of the parameter q above a given limit, these spike train metrics give a smaller distance between a real spike train and a null spike train, than the distance between any two real spike trains from the same cell! This result sheds some light into the limit values that must be used for the parameter q. That is, the values for q must not be too high, otherwise the error will be smaller for the distance between a spike train and a null spike train than between real spike trains. This drawback is particularly relevant for tuning a retina model, since the obtained parameters in the optimization process can lead the model to not fire any spike at all, the opposite from what is intended!

The computational implementation of the spike train distance metric (d^2) is straightforward. The graphic in Fig. 5.8 displays the minima and maxima evolution of this metric as a function of the kernel's decaying rate (τ), by comparing the different trials from the salamander ON-type RGC. One difficulty of this metric is the choice of the right value for τ, due to the fact that it scales the kernel in Eq. (5.28). For $\tau \ll T_s$ the discrete kernel, given in Eq. (5.28), will always be equal to the discrete sequence

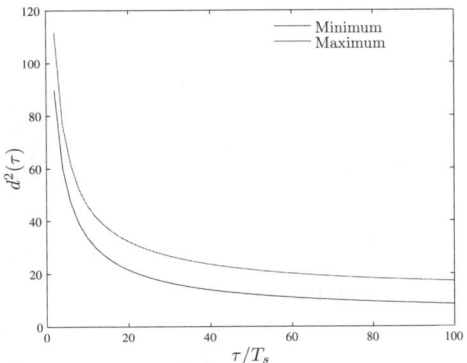

Fig. 5.8 Minimum and maximum values for the distance metric for the salamander ON-type RGC responses.

$[1\ 0\ 0\ \ldots]$; convolution will leave the discrete neuronal response function $\rho[n]$ unchanged, while the value of the numeric integration is divided by τ (see Eq. (5.29)), leading to erroneous values that can increase unboundedly for small τ. Thus, τ must be several orders of magnitude bigger than the sampling period T_s (1 to 3 orders), so that the numerical integration expressed by Eq. (5.29) gives meaningful results. For values of $\tau \to \infty$ the values of the metric should tend to Eq. (5.20), but this tendency is very slow, and for big values of τ the kernel sequence will be quite long; the calculations will become lengthy, and the memory requirements can become prohibitive.

5.5 Spike Events Metrics

The *spike events metric*, $d^{events}(\cdot)$, has been proposed to measure the distance between two sets of spikes trains: a reference and an estimated set of spike trains [Keat et al. (2001)]. After the definition of the firing events in each set, this metric employs the matching principle used in [Victor and Purpura (1996)] to match the events.

This metric is intended to compare sets of spike trains corresponding to the responses of a neuron to the same stimulus – the set of reference trials, with a set of spike trains predicted by a model, for example. The comparison is made in terms of firing events, where a firing event is obtained by grouping identical bursts of spikes within the spike trains. Figure 5.9

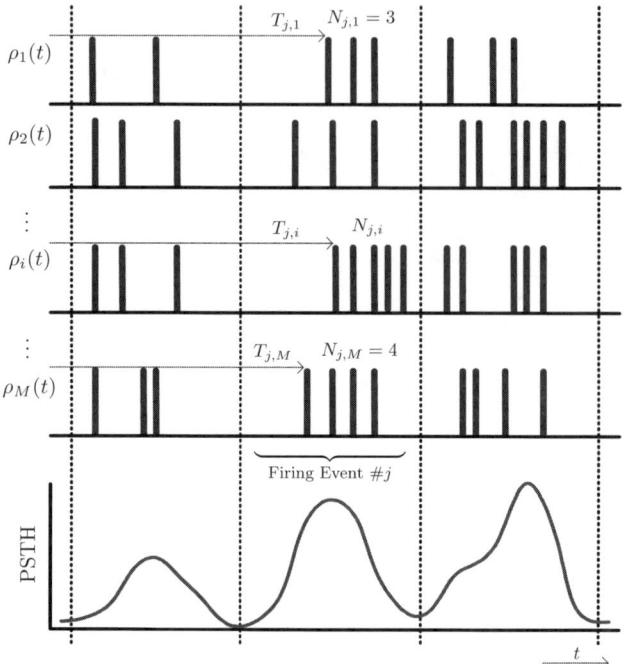

Fig. 5.9 Parsing a set of spike trains into firing events.

represents a set of M spike trains with event classification.

The events metric is based on the assumption that the neuron responds, within a certain range, with similar spike trains to the same stimulus. These spike trains are characterized by regions where not even a single spike is fired, followed by a burst of spikes – a spike event – in response to stimuli (see Fig. 3.3). Moreover, it was observed, particularly in the case of the retina, that these trains are reproducible from trial to trial, not only in terms of the time occurrence of the spikes but also regarding the number of spikes [Reinagel (2001)]. The spike train metrics previously described were found to be unsuited to measure the reliability of such trains. Therefore, the spike events metric takes into account the time occurrence of spikes, the number of spikes, the variation in the time occurrence, and the variation in the number of spikes for each event.

The computation of the spike events metric consists of parsing the spike trains into firing events (see Fig. 5.9), and matching events between the two sets. The firing events correspond to bursts of spikes delimited by regions

Algorithm 5.3 Gaussian filter width σ, $GaussFilterWidth$

1: $T \leftarrow$ sampling period
2: $M \leftarrow$ number of spike trains
3: $\boldsymbol{\rho}_{M,:} \leftarrow$ matrix whose rows are the discretized reference spike trains
 {Accumulate the cross-correlation between all trials of spikes for a maximum lag l_{max}:}
4: **for** $i = 1$ to $M - 1$ **do** {For all trials in $\boldsymbol{\rho}$}
5:    **for** $j = i$ to M **do** {For trials in $\boldsymbol{\rho}$ not equal to $\boldsymbol{\rho}_{i,:}$}
6:       $\mathbf{c} = \mathbf{c} + \mathrm{corr}(\boldsymbol{\rho}_{i,:}, \boldsymbol{\rho}_{j,:})$
7:    **end for**
8: **end for** {$\mathbf{c}$ as dimensions $2l_{max} + 1$}

   {Take into account only the main lobe of the histogram $\mathbf{c}$}
9: $m \leftarrow \mathbf{c}_1$
10: **for** $i = 1$ to l_{max} **do**
11:    **if** $m < \mathbf{c}_i$ **then**
12:       $l = i$; $m = \mathbf{c}_i$
13:    **end if**
14: **end for**
15: $\mathbf{c} \leftarrow \mathbf{c}_l$ to $\mathbf{c}_{2l+1}$

   {Calculate the mean of the histogram: μ}
16: $\zeta \leftarrow 0$; $n \leftarrow 0$
17: **for** $i = 1$ to $2l + 1$ **do**
18:    $\zeta = \zeta + \mathbf{c}_i(i - (l+1))T$
19:    $n = n + \mathbf{c}_i$
20: **end for**
21: $\mu = \frac{\zeta}{n}$

   {Variance calculation: σ^2}
22: $\epsilon \leftarrow 0$
23: **for** $i = 1$ to $2l + 1$ **do**
24:    $\epsilon = \epsilon + (\mathbf{c}_i(i - (l+1))T)^2$
25: **end for**
26: $\sigma^2 = \frac{\epsilon}{n} - \mu^2$
27: $\sigma = \sqrt{\sigma^2}$ //Standard deviation

   {Finally, taking into account that the contribution to the interspike interval is due to spikes from both trials}
28: $\sigma = \frac{\sigma}{\sqrt{2}}$
29: **return** σ

where the firing rate is nearly zero. Thus, to compute the boundary values between firing events, a PSTH is obtained from the set of spike trains. This PSTH is usually smoothed by convolving it with a Gaussian filter (see Eq. (3.29)) whose width, σ, is defined as the time scale of modulations in the firing rate. Specifically, the value of σ is obtained by adjusting a Gaussian function to the histogram of time differences between all pairs of spikes trains within a trial. The interspike interval histogram can be obtained by correlating all pairs of trials from the set (see Sec. 3.2.4). The value of σ is made equal to the width of the Gaussian curve fitted to the interspike time histogram divided by $\sqrt{2}$, because both spikes from the two spike trains contribute to the interspike time jitter [Berry et al. (1997b)]. Algorithm 5.3 details the calculation of the width of the smoothing filter.

Figure 5.10(a) shows a segment with a 2 s duration of a set of 13 trials of a rabbit RGC when excited by ON-OFF type stimuli (see Fig. 3.5(a)), and Fig. 5.10(b) shows the real and the smoothed PSTH. The minima of the smoothed PSTH are calculated and they correspond to the locations where the firing event boundaries should be positioned. The spikes between a pair of such minima are considered part of the same firing event. Since not all minima are equal to zero, a rule must be used to distinguish between real firing event boundaries and just a local decrease in the firing rate. Therefore, a given minimum m_i, between the maxima M_i and M_{i+1} in the PSTH, is considered to be a firing event boundary if it is significantly smaller than its neighboring maxima according to the rule:

$$\frac{\sqrt{M_i M_{i+1}}}{m_i} \geq \phi. \tag{5.30}$$

If the condition in Eq. (5.30) is verified, then m_i is considered a true minimum that identifies a boundary between two adjacent firing events; otherwise, the two initial firing events are, in fact, a single event. The original values used for ϕ belong to the set $\phi \in \{1.5, 3\}$, but experimental results have shown that the chosen value does not have too much influence in the result [Keat et al. (2001)]. Figure 5.10(b) also shows the firing events obtained by the application of the criteria decision expressed by Eq. (5.30) with $\phi = 3$. After being defined, each firing event j from every trial i is characterized by two numbers: the time of the first spike, $T_{j,i}$; and by the number of spikes, $N_{j,i}$ (see Fig. 5.9). Four quantities are calculated for each firing event, in order to generate the correspondence between the firing events of the two sets of trials. For the event j from a set comprising M trials (refer to Fig. 5.9) the quantities of interest are:

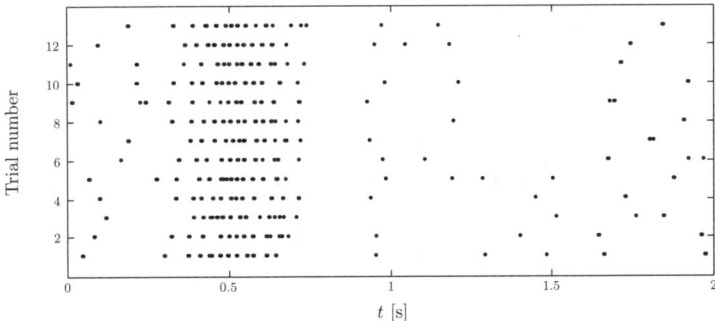

(a) Trains of spikes from a rabbit ON-type cell. Each spike is represented by a dot.

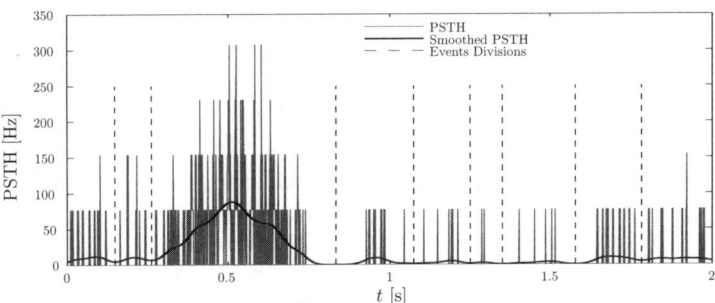

(b) Real and smoothed PSTH with the firing events divisions.

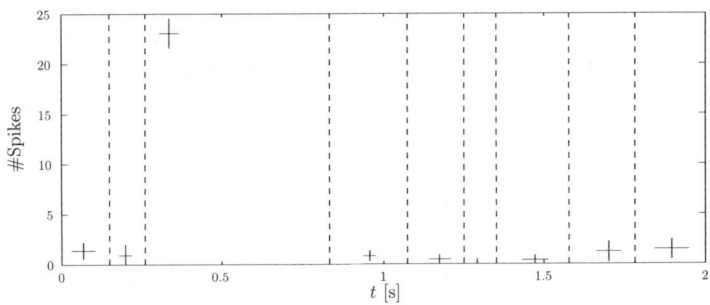

(c) Firing events characterization: the values of the cross center (x, y) correspond to the mean time of first spike, T, and to mean number of spikes, N, respectively; the cross width represents the standard deviation of the time occurrence of the first spike, δT, while its height is the standard deviation of the number of spikes, δN.

Fig. 5.10 Characterization of 13 observed trials into firing events.

- the average across trials of the time occurrence of the first spike within the firing event:

$$T_j = \frac{1}{M} \sum_{i=1}^{M} T_{ji} ; \qquad (5.31)$$

- the average number of spikes across trials in the event:

$$N_j = \frac{1}{M} \sum_{i=1}^{M} N_{ji} ; \qquad (5.32)$$

- the standard deviation of the time occurrence of the first spike, in an event, across trials, which measures the time jitter of the first spike in the event:

$$\delta T_j = \sqrt{\frac{1}{M-1} \sum_{i=1}^{M} (T_{ji} - T_j)^2} ; \qquad (5.33)$$

- the standard deviation of the number of spikes across the trials within a firing event, which measures the deviation from the mean spike number:

$$\delta N_j = \sqrt{\frac{1}{M-1} \sum_{i=1}^{M} (N_{ji} - N_j)^2} . \qquad (5.34)$$

This procedure maps the set of spike trains into a sequence of firing events – a train of events – where each event is characterized by the four quantities: T, N, δT, and δN. Figure 5.10(c) shows the characterization of the firing events from the set of spike trains of the rabbit ON-type cell of Fig. 5.10(a). The algorithm described in Algorithm 5.4 describes the steps followed to parse a set of trials into a sequence of firing events. The next step in computing the spike event metric is to match the firing events from the observed retina responses with the predicted neural responses.

If we order the firing events of the observed set according to index i and the ones from the estimated set according to index j, the overall expression for the error between the two event trains is given by

$$d^{event} = e_T E_T + e_N E_N + e_{\delta T} E_{\delta T} + e_{\delta N} E_{\delta N} - e_M E_M. \qquad (5.35)$$

This distance measure includes five sources of error between the observed and predicted spike trains, namely: E_T, E_N, $E_{\delta T}$, $E_{\delta N}$, E_M. The error

Algorithm 5.4 Parsing spike trains to events, $Spike2Events$

1: $\rho \leftarrow$ matrix whose rows are the spike trains
  {Compute:
  - The firing rate $\mathbf{r}$ from $\rho_{M,N}$
  - Smooth $\mathbf{r}$ with a Gaussian filter ($\sigma \leftarrow GaussFilterWidth(\rho)$)
  - Compute the minima and maxima of the rate r; every minimum m_i is delimited by the maxima M_i and M_{i+1}, such that the following conditions are true:
    - $m_i \leq r(k)$, for all $\forall_{N_i < k < N_{i+1}}$, where N_i is the timestamp of the maxima M_i
    - $M_i \geq r(k)$, for all $\forall_{n_{i-1} < k < n_i}$, where n_i is the timestamp of the minima m_i
    - $M_i \geq m_i \quad \wedge \quad M_{i+1} \geq m_i$
  }
2: $(\mathbf{m}, \mathbf{n}) \leftarrow$ vectors of values and time stamps of the minima of rate r, i.e. $\mathbf{m} = r(\mathbf{n})$
3: $(\mathbf{M}, \mathbf{N}) \leftarrow$ vectors of values and time stamps of the maxima of rate r, i.e., $\mathbf{M} = r(\mathbf{N})$ {Remove fake minima according to criteria of (5.30) or that leads to an event with no spikes}
4: **for all** $m_i \in \mathbf{m}$ **do**
5:   **if** $\sqrt{M_i M_{i+1}}/m_i < \phi$ **then**
6:     Remove minimum m_i and its index n_i from the vectors $\mathbf{m}$ and $\mathbf{n}$, respectively
7:     Remove the smallest of the maxima $\{M_i, M_{i+1}\}$ and its index from the vectors $\mathbf{M}$ and $\mathbf{N}$
8:   **end if**
9: **end for**
  {Compute event statistics}
10: **for all** $m_i \in \mathbf{m}$ **do**
11:   $\mathcal{T}_i \leftarrow$ average time stamp of the first spike in the interval $[n_{i-1} \ n_i]$
12:   $\mathcal{N}_i \leftarrow$ average number of spikes in the interval $[n_{i-1} \ n_i]$
13:   $\delta\mathcal{T}_i \leftarrow$ standard deviation in the time of first spike in $[n_{i-1} \ n_i]$
14:   $\delta\mathcal{N}_i \leftarrow$ standard deviation in the number of spikes in $[n_{i-1} \ n_i]$
15: **end for**
16: **return** $\mathcal{T}, \mathcal{N}, \delta\mathcal{T}, \delta\mathcal{N}$

terms in Eq. (5.35) are defined as:

$$E_T = \sum_{\substack{\text{matched} \\ \text{event pairs } (i,j)}} |T_i - \hat{T}_j|; \tag{5.36}$$

$$E_N = \sum_{\substack{\text{matched} \\ \text{event pairs } (i,j)}} |N_i - \hat{N}_j| + \sum_{\substack{\text{unmatched} \\ \text{events } i}} N_i + \sum_{\substack{\text{unmatched} \\ \text{events } j}} \hat{N}_j; \tag{5.37}$$

$$E_{\delta T} = \sum_{\substack{\text{matched} \\ \text{event pairs } (i,j)}} |\delta T_i - \widehat{\delta T}_j|; \tag{5.38}$$

$$E_{\delta N} = \sum_{\substack{\text{matched} \\ \text{event pairs } (i,j)}} |\delta N_i - \widehat{\delta N}_j|; \tag{5.39}$$

$$E_M = \sum_{\substack{\text{matched} \\ \text{event pairs } (i,j)}} 1. \tag{5.40}$$

The hats over the quantities in the expressions of Eq. (5.36) – Eq. (5.40) refer to the values obtained from the firing events estimated by the model, while the quantities without a hat are related to the observed firing events from a real cell. The expression in Eq. (5.40) indicates that the matching of events between sequences is rewarded by adding a negative contribution to the error in Eq. (5.35).

The weights for each error component in Eq. (5.35) are obtained from

$$
\begin{aligned}
e_T &= \frac{1}{E\{\delta T\}}; \\
e_N &= \frac{1}{E\{\delta N\}}; \\
e_{\delta T} &= \frac{1}{2E\{\delta T\}}; \\
e_{\delta N} &= \frac{1}{2E\{\delta N\}}; \\
e_M &= 2;
\end{aligned}
\tag{5.41}
$$

where the averages are calculated across all events in the event sequence. That is, if we have a total of Q events in the reference set of trials, these values are

$$E\{\delta T\} = \frac{1}{Q}\sum_{j=1}^{Q} \delta T_j \; ; \quad E\{\delta N\} = \frac{1}{Q}\sum_{j=1}^{Q} \delta N_j. \tag{5.42}$$

The mean of the standard deviation of the first spike occurrence within the trials ($E\{\delta T\}$) is used to scale the error related to the time jitter differences between the neurons and the predicted trials, while the mean of the standard deviation of the number of spikes ($E\{\delta N\}$) plays the same role by scaling the differences in the number of spikes. The coefficients $e_{\delta T}$ and $e_{\delta N}$ are equal to half of e_T and e_N, respectively, meaning that the spike occurrence time and the spike number are twice as important to the error measurement than their variation. The constant value for e_M gives a negative contribution to the error, rewarding the matching of two events.

A procedure to match the events from the real and predicted firing events must be devised. To match the events, a recursive procedure, similar to the one described to match different spike trains in the spike time metric (and in the interspike interval metric), can be employed. To obtain a recursive procedure, we can rely on the following restriction for the alignment of two firing events sequences: two events in one train cannot be matched to two events in the other train that occur in reverse order. We represent the sequence of firing events from the reference trials by R, which possesses n_1 events, and the sequence of firing events from estimated trials by $\hat{R}$, which possesses n_2 events. In matching the event i from R with the event j from $\hat{R}$, one of three cases can occur: *i*) the last analyzed event in R is unmatched; *ii*) the last analyzed event in $\hat{R}$ is unmatched; or *iii*) the last events in R and $\hat{R}$ match each other. If $d_{i,j}^{events}$ represents the error incurred in matching the first i events of R with the first j events of $\hat{R}$, taking into account the previous restrictions, the events can be matched iteratively. Each one of the previous three possibilities leads to three different values for the matching error, and the smallest value should be chosen to match event i with event j according to

$$d_{i,j}^{events} = \min\left\{d_{i-1,j}^{events} + e_N N_i; d_{i,j-1}^{events} + e_N \hat{N}_j; d_{i-1,j-1}^{events} + M_{i,j}\right\}, \quad (5.43)$$

where the cost $M_{i,j}$ for matching events i and j is obtained from

$$M_{i,j} = e_T|T_i - \hat{T}_j| + e_N|N_i - \hat{N}_j| + e_{\delta T}|\delta T_i - \widehat{\delta T}_j| + e_{\delta N}|\delta N_i - \widehat{\delta N}_j| - e_M. \quad (5.44)$$

Based on Eq. (5.43), we can initiate a recursive procedure that will iteratively match the events from the two trains. By starting with $d_{0,0}^{events} = 0$, we can fill a matrix for the match of the first i events of one train to the first j events of the other train by recursively calculating $d_{i,j}^{events}$. The last element in the diagonal contains the total error given by Eq. (5.35). Algorithm 5.5 presents the sequence of operations to match two trains of events.

Algorithm 5.5 Spike events metric, d^{events}
1: $\boldsymbol{\rho}(t) \leftarrow$ set of observed spike trains
2: $\hat{\boldsymbol{\rho}}(t) \leftarrow$ set of estimated spike trains
   {Compute event statistics, as described in Algorithm 5.4; the function results in four vectors for each set of trials}
3: $\{\boldsymbol{T}, \boldsymbol{N}, \boldsymbol{\delta T}, \boldsymbol{\delta N}\} \leftarrow Spike2Events(\boldsymbol{\rho})$
4: $\{\widehat{\boldsymbol{T}}, \widehat{\boldsymbol{N}}, \widehat{\boldsymbol{\delta T}}, \widehat{\boldsymbol{\delta N}}\} \leftarrow Spike2Events(\hat{\boldsymbol{\rho}})$
5: $n_1 \leftarrow$ number of events in $\boldsymbol{\rho}(t)$, i.e., number of elements in the vectors $\boldsymbol{T}$ (and in the other statistics' vector)
6: $n_2 \leftarrow$ number of events in $\hat{\boldsymbol{\rho}}(t)$, i.e., number of elements in the vectors $\widehat{\boldsymbol{T}}$ (like in the other statistics)
   {Compute weights for each penalty as stated in (5.41)}
7: $e_T \leftarrow 1/E\{\boldsymbol{\delta T}\}; e_N \leftarrow 1/E\{\boldsymbol{\delta N}\};$
8: $e_{\delta T} \leftarrow 1/(2E\{\boldsymbol{\delta T}\}); e_{\delta N} \leftarrow 1/(2E\{\boldsymbol{\delta N}\}); e_M \leftarrow 2$
   {Compute the distance between the first i events of $\boldsymbol{\rho}(t)$ and the first j events of $\hat{\boldsymbol{\rho}}(t)$}
9: $\mathbf{D} \leftarrow$ matrix of size $[(n_1+1) \times (n_2+1)]$ where
   - first row: $\mathbf{D}_{0,:} = [0, e_N \widehat{\boldsymbol{T}}^T]$
   - first column: $\mathbf{D}_{:,0} = [0, e_N \boldsymbol{T}]^T$
10: **for** $i = 1$ to n_1 **do** {For all events of $\boldsymbol{\rho}$}
11:    **for** $j = 1$ to n_2 **do** {For all events of $\hat{\boldsymbol{\rho}}$}
12:       $M = e_T|\boldsymbol{T}_i - \widehat{\boldsymbol{T}}_j| + e_N|\boldsymbol{N}_i - \widehat{\boldsymbol{N}}_j| + e_{\delta T}|\boldsymbol{\delta T}_i - \widehat{\boldsymbol{\delta T}}_j| + e_{\delta N}|\boldsymbol{\delta N}_i - \widehat{\boldsymbol{\delta N}}_j| - e_M$
13:       $\mathbf{D}_{i,j} = \min\left\{\mathbf{D}_{i-1,j} + e_N N_i; \mathbf{D}_{i,j-1} + e_N \hat{N}_j, \mathbf{D}_{i-1,j-1} + M\right\}$
14:    **end for**
15: **end for**
16: **return** $d^{events} \leftarrow \mathbf{D}_{n_1,n_2}$

As pointed out in [Keat *et al.* (2001)], not all terms of the matrix need to be calculated, since for two events far apart in time, the error $M_{i,j}$ is so large that the two events will never match. Specifically, events that occur farther apart than the following condition need not to be calculated:

$$|T_i - \hat{T}_j| > \frac{1}{e_T}\left(2e_N N_{max} + e_M\right), \quad (5.45)$$

where N_{max} is the largest number of spikes in an event from the event trains under comparison.

5.5.1 Spike Events Metric Analysis

The spike events metric has the particularity that, every time two events match with each other, there is a negative contribution – a compensation – to the overall error value. However, the application of this metric requires several trials of the response of a given cell for the same stimulus in order to define the events. Moreover, this metric violates the condition in Eq. (5.1) to define a metric function: it can give a negative distance between two sets of spike trains. Nevertheless, this problem could be solved by adding an offset to the origin by knowing the number of firing events in the reference set. The value to add to the firing events error in Eq. (5.35) would be Me_M, where M is the number of events present in the reference data. Although this leads to the desired situation that if the two sets are equal their distance will be zero, as Eq. (5.1) states, with this new term the property of Eq. (5.2) is not respected. Moreover, using the statistics of the reference spike train to compute the weights in Eq. (5.41) this metric does not fulfill the condition in Eq. (5.2).

The main difficulty in the implementation of this metric is in the delimitation of the events, what passes through the calculation of the minima of the smoothed PSTH that must appear between two maxima. Furthermore, since we can have a real maxima, or simply a transition to, or from, a constant firing rate, extra care must be taken to validate a minimum as a true one.

Firing event divisions must be positioned at the smoothed PSTH minima. Thus, its minima were calculated and classified as event divisions. This classification avoids barely pronounced minima being considered a firing event division. A minima is classified as a firing event division if it obeys the condition given in Eq. (5.30). After the definition of the spike events, the quantities in Eq. (5.36)– Eq. (5.40) are calculated, and the value for the error is obtained by the successive application of Eq. (5.43) (see Fig. 5.10).

As we have previously remarked, $d^{events}(\cdot)$ does not possess the metrics' property in Eq. (5.1); by comparing the set of spike trains with itself, the error is not zero but equal to $-2 \times M$, where M is the number of events. This is because of the negative term in Eq. (5.35) and all events are perfectly matched between the two sets of spike trains.

5.6 Tuning and Assessment of Retina Models

Let us perform an experimental analysis of the retina functional models described in Chap. 4 using the different metrics described in this chapter. This analysis is important not only to assess the retina models, but also to tune them. To analyze the models, they must be tuned first according to one metric, and after this first procedure, the performance can be evaluated. Different data must be used to tune and to assess the models. Thus, it is usual to divide the available data into training and test data sets, with the test data set smaller than the training set. In order to evaluate the metrics and the models, the retina data used for the tuning and evaluation are first characterized using the same metrics.

The retina data used were obtained by sampling the responses of stimulated, real animal RGCs. Two different sets were used: *i)* a data set gathered for a full-field flash stimuli; and *ii)* a data set corresponding to full-field stimuli, but with stimuli that flickered randomly with an intensity value drawn from a Gaussian distribution. Data for the flash stimulus were provided by the CORTIVIS consortium project [Project CORTIVIS (2006)], and were collected at the facilities of the University Miguel Hernandez (Alicante, Spain). This data was obtained through extracellular recordings from ganglion cell populations in isolated superfused albino rabbit (Oryctolagus cuniculus) retina using a rectangular array of 100 electrodes with 1.5 mm long (see Sec. 3.2.1 for more details about the stimulation and recording apparatus). The stimuli corresponded to full-field flash stimuli of the ON-OFF type, (see Figure 3.7(a)), where the light intensity was kept to its maximum value for 300 ms (RGB values: $(255, 255, 255)$), and switched off to its minimum intensity (RGB values: $(0, 0, 0)$) for 1700 ms, thus providing maximum contrast. Figure 3.5 shows a fraction of the stimulus and the responses to the stimulus.

The second set of data used was provided by the Meister Lab [Meister (2007)]. It was gathered using an apparatus for data acquisition described in [Meister *et al.* (1994)]. The stimuli consisted of a spatially uniform illumination that flickered randomly with an intensity value drawn from a Gaussian distribution. The full-field intensity was obtained by sampling every 30 ms a normal distribution with standard deviation equal to 35% of the mean level intensity with the value 4 mW/m^2 [Keat *et al.* (2001)]. Figure 3.7(b) illustrates the spatiotemporal characteristics of such stimuli, and Fig. 3.6 shows the stimuli and part of the respective ganglion cells' responses for the experimental data used, corresponding to a salamander

ON-type RGC.

One of the problems encountered while optimizing a function or tuning a model with a considerable number of parameters is the absence of information about the shape of the function to minimize in the parameter space. In the case of tuning the retina models, this can be due not only to the number of parameters involved, but also to the intrinsic nature of the metrics. Therefore, nonlinear optimization [Flannery et al. (2002)] must be applied. The results presented next were obtained using unconstrained nonlinear optimization with the Nealder-Mead simplex algorithm, complemented with a simulated annealing scheme based on the method described in [Efstratiadis and Koutsoyiannis (2002)]. The RGC responses were discretized with a sampling period of $T_s = 1$ ms.

The models' parameters were tuned to optimize their responses with respect to the NMSE, d^{time} and d^{events} metrics. The spike time metric was employed with a value for the parameter $q = 0.2$ (ms)$^{-1}$, such that two spikes in different trials that were more than $q/2$ apart were considered not to be related. Thus, the spike precision occurrence within a firing was $1/q = 5$ ms. Generally, the tuning of the models with the spike time metric gave an error for this same metric equal to the number of spikes present in the trials, meaning that all spikes were created in the distance measure process. Next, we will summarize the analysis for each functional model presented in Chap. 4.

5.6.1 Deterministic Model

The deterministic model performs well when tuned and assessed with the NMSE metric for the ON-OFF-type RGC responses, mainly because this model is well adapted to this type of stimulus. Figure 5.11(a) shows the PSTH of both the RGC and the model responses to the ON-OFF stimulus. Tuning the deterministic model with the spike events metric gives a larger error when assessed with the NMSE metric than when tuned with this same metric, while the error for the spike events metric slightly increases as well (see Fig. 5.16). This means that the model is incapable of reproducing the events, as Fig. 5.11(b) shows, which is demanded by the events metric.

The deterministic model is unable to reproduce the RGC behavior to the white noise stimulus irrespective of the training metric, as the displays in Fig. 5.12 exemplify. When tuned with the NMSE metric for the white noise stimulus, the model is only capable of generating a firing rate equal to the mean firing rate produced by the cell. Errors in Fig. 5.17 for the NMSE

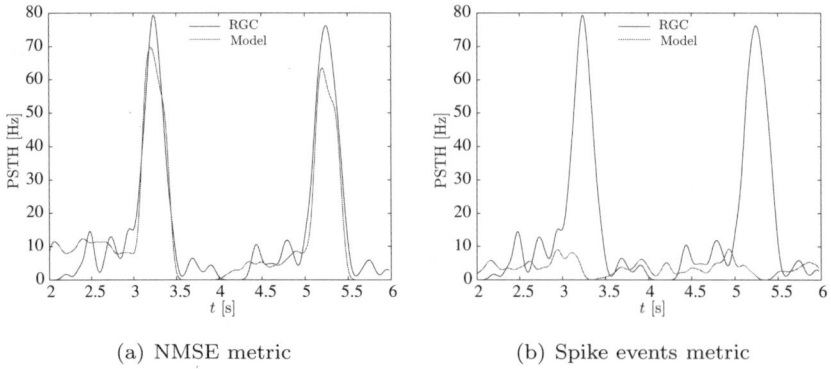

(a) NMSE metric (b) Spike events metric

Fig. 5.11 Deterministic model responses, when tuned with (a) the NMSE and with the (b) spike events metrics, to the flash stimulus.

metric show to be small compared regarding the ON-OFF data, but this happens because the white noise data are characterized by an intrinsic high NMSE value (so the ratio becomes small).

For the spike events metric, the errors are much higher than the errors characterizing the data. The deterministic model cannot predict the events nor place them in the correct position. This metric is not suitable to optimize this model relative to the white-noise data, or the model is not well adapted to model the RGC response for this type of stimulus.

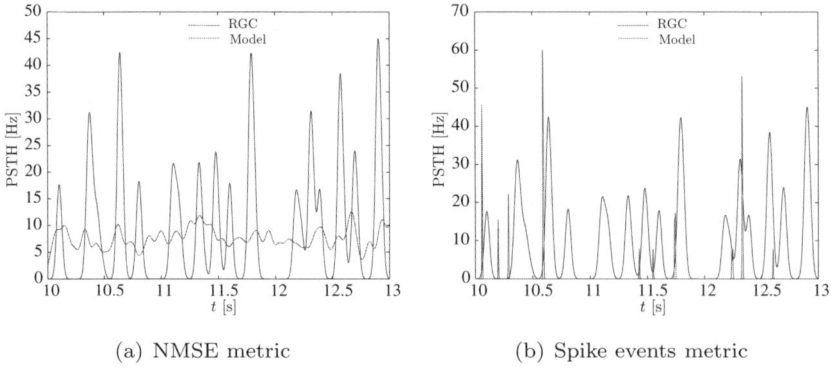

(a) NMSE metric (b) Spike events metric

Fig. 5.12 Deterministic model responses, when tuned with (a) the NMSE and with the (b) spike events metrics, to the white noise stimulus.

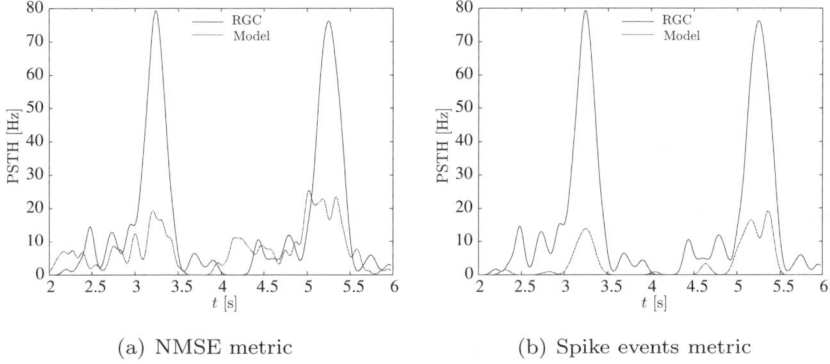

Fig. 5.13 Stochastic model responses when tuned with (a) the NMSE and (b) spike events metrics to the flash stimulus.

5.6.2 Stochastic Model

The stochastic model was initialized with two different sets of parameters from [Keat et al. (2001)], depending on whether the ON-type RGC data were from a rabbit or salamander. By tuning and assessing the model with the NMSE metric for the flash data, the errors obtained are higher than those that characterize the test data, as Fig. 5.16 shows. The model follows the main lobes of the observed firing rate imprecisely (see Fig. 5.13(a)), giving large errors for all metrics.

The use of the spike events metric for tuning the model gives the best qualitative and quantitative results (see Fig. 5.13(b) and Fig. 5.16). For the white noise stimulus, the stochastic model produces the best results when tuned with the spike events metric, as can be seen from Fig. 5.17. Figure 5.14 shows the qualitative results obtained by applying the NMSE metric and the firing event metric to the white noise data.

From these results, it can be stated that the stochastic model gives the best results when tuned and assessed with the spike events metric.

5.6.3 White Noise Model

The white noise model is directly tuned from the training data, as it exempts the use of an optimization process with a specific metric. Figure 4.12 shows the qualitative results of the model when applied to the flash and to the white noise stimuli. Comparing the results obtained with this model for the different types of data, we can see from Fig. 5.16 and Fig. 5.17 that this

192 Bioelectronic Vision: Retina Models, Evaluation Metrics, and System Design

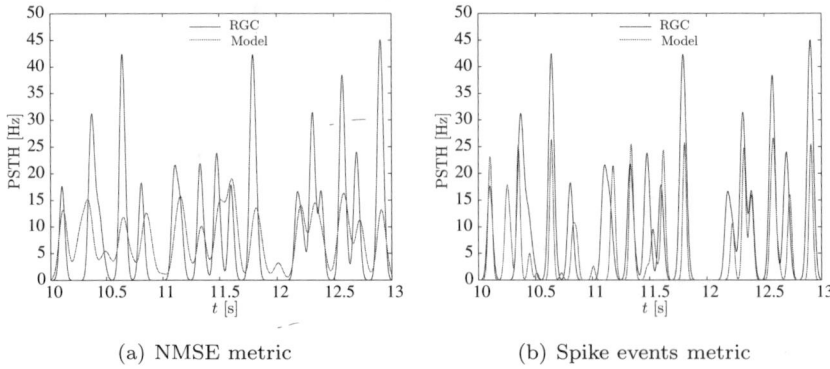

Fig. 5.14 Responses of the stochastic model to the white noise stimulus, when tuned with (a) the NMSE and (b) spike event metrics.

model is the most versatile among those analyzed, as it shows the lowest errors for the different metrics and for the two stimulus scenarios.

From the analyzed models, and for the two types of stimuli used, we can say that the deterministic model performs better for the flash-like stimulus when tuned with a firing rate metric, like the NMSE metric, while the responses to the white noise stimulus are better described by the stochastic model when tuned with the spike events metric. As a matter of fact, these were the metrics originally employed in the development phase of these models. So, we can state that a given model is tied to the metric or, equivalently, to the neural code perspective followed in its development. Moreover, the white noise model, which is not tied to any particular met-

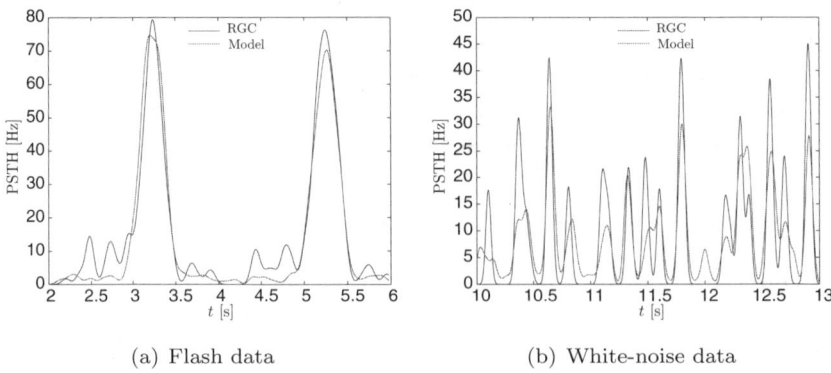

Fig. 5.15 White noise model responses to the (a) flash and to the (b) white noise stimuli.

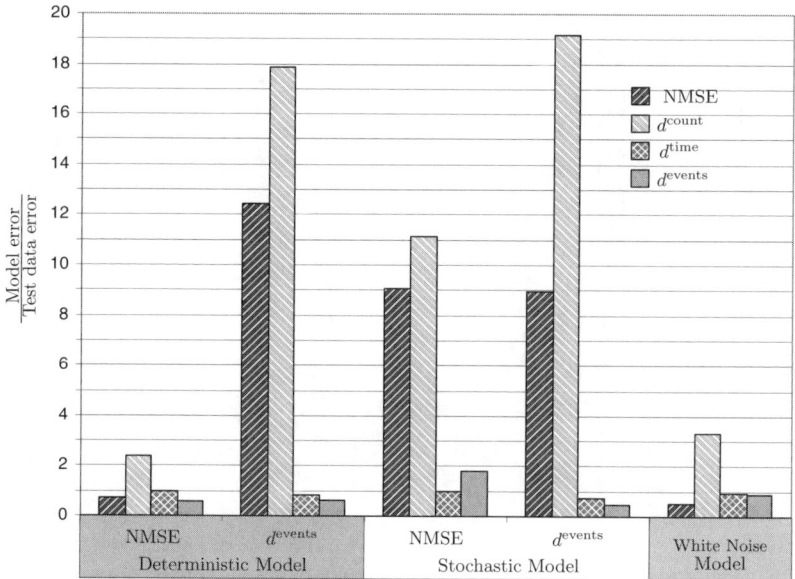

Fig. 5.16 Models errors for the ON-OFF stimulus (horizontal axis: training method/model; legend: assessment metric).

ric, has shown to be the most adjustable, having a similar performance irrespective of the stimulus type.

5.7 Conclusions and Further Reading

The neural code metrics presented in this chapter are scattered and embedded through the specialized literature about the subject. However, it is important to have a framework for comparison and evaluation of neural models, and of retina models in particular. This is the principal goal of the previous sections.

The analyzed models are representative of three classes of functional models described in Chap. 4, and at the end of this chapter a main question still remains: among the available models which one models the retina best? This question has been shown not to have a simple and single answer, mainly because there is not a universal retina model applicable in every situation, so several issues must be considered.

With these representative models on hand, another question is raised: which error measure, or metric, should be used to evaluate the models?

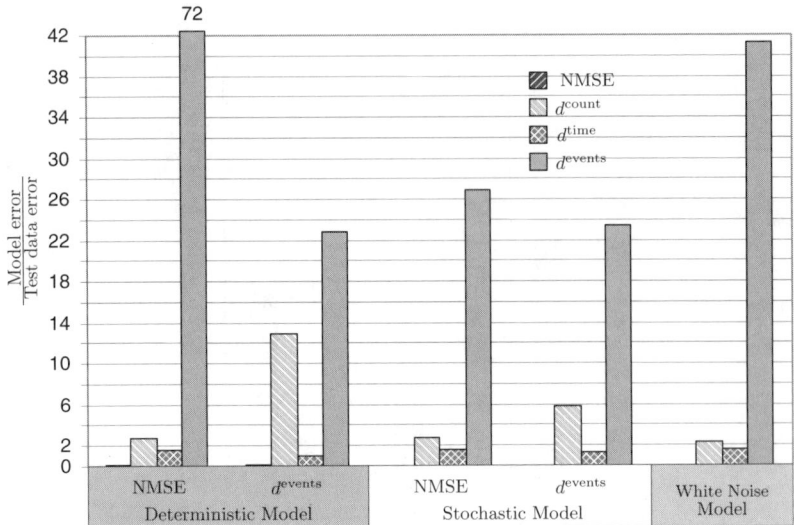

Fig. 5.17 Models errors for the white noise stimulus (horizontal axis: training method/model; legend: assessment metric).

This fundamental question also does not have a unique and straight answer. The response to this question is equivalent to determining the relevant characteristics of the spike trains sent by the peripheral senses, like the retina, used by the human brain to extract the sensed information (which is the same code that the brain uses to send information to the members). In particular, the mapping between a visual image and the spike trains sent by the retina to the brain involves two concepts, termed *ambiguity* and *variability*. The ambiguity concept expresses the fact that two different images can elicit similar sets of spike trains, while variability expresses the fact that the same image can evoke rather different spike trains. We have gathered the most common metrics used in the neuroscience field to develop retina models, and also other metrics that scientists believe to be used in the neural code. Namely, we have presented firing rate metrics like the MSE, the NMSE, and the %VAF metric [Berry II and Meister (1998); Westwick and Kearney (2003); Pillow *et al.* (2005)]; in the spike train metrics class, we have discussed the spike time metric, the spike interval metric [Victor (2005); Victor and Purpura (1996, 1997)], and the spike distance metric [van Rossum (2001)]; and in the events class, we have discussed the spike events metric [Keat *et al.* (2001)].

In assessing the models, we have used two different types of visual stimuli. The deterministic stimulus represents better geometrical images, such as artificial landscapes like buildings or text, while the random stimuli possess characteristics similar to natural landscapes, like a forest [Dong and Atick (1995)].

The metrics collected measure different characteristics of the neural code, and so the establishment of a criteria in which they can be ranked is hard to find. As stressed by several researchers in the field, the firing rate metric does not seem to be the most likely metric used by the neural code. If this is definitively confirmed, the spike train and spike events metrics should be used to develop and optimize the retina models.

As stressed by the performance analysis, the spike time metric was shown to be useless in the model tuning. We can advance two reasons for this: the metric is incapable due to its nature, explored in Section 5.4.4, of being able to optimize a model since it has a pronounced minimum for the case when no spike is fired that the optimization process always converges to, or the analyzed models are incapable of reproducing the retina response using this metric criteria. Notwithstanding, this metric is very useful to measure the precision of a spike train. The choice of one of the spike train metrics, or the spike events metric, is tied to the biological interpretation of the neural function. Finding a general and universal metric to measure and evaluate the neural code still remains open, and it may remain so for the time to come.

The main conclusion that can be drawn from the analyzed models is that the deterministic and stochastic models are tied to the metric and to the type of stimulus they model [Martins and Sousa (2005)]. The white noise model is the most adaptable, producing good maps between the stimuli and the respective responses. A priori, this model would be expected to be completely adaptable to a given stimulus/response pair based on analyzing its structure. Nevertheless, this model cannot be tailored to respond to one type of stimulus and be expected to produce the same good results when applied to other different types of stimuli.

In this chapter, we hope to have shown that a universal retina model, capable of describing meaningful responses for all types of visual stimuli existent in everyday life, is far from being achieved. For now, the efforts are concentrated in modeling the retina response to isolated categories of stimuli. In the next chapter, we will describe a bioelectronic prosthesis based on one of the analyzed models.

Exercises

5.1. Generate two sets of several spike trains using a spike generator, and apply the firing rate metrics to measure the distances between the firing rate obtained with these sets. Use different parameters to generate the spike trains sets and observe the results. Try the same procedure with real spike trains.

Suggestion: Use the procedure described in Exercise 3.3.

5.2. Generate two spike trains with a Poisson generator and compare them using the spike train metrics. Repeat this process with a integrate-and-fire generator and evaluate the results.

5.3. Using the two following discrete spike trains, where a 1 means that a spike was fired in that time bin and a 0 indicates that there was no neural activity,

$$\rho_1: 1\ 0\ 1\ 0\ 0\ 1\ 1$$
$$\rho_2: 1\ 0\ 0\ 1\ 0\ 1$$

construct the cost matrices considering that the sampling period is $T_s = 1$ ms, and $q = 0.9$ $(ms)^{-1}$, for:

- **5.3.1** the spike time metric, $d^{time}(\cdot)$, and obtain the final spike time distance between the trains;
- **5.3.2** the interspike interval metric, $d_q^{inter}(\cdot)$. What is the interspike interval distance between the trains?

5.4. For the discrete spike trains below, where the only difference is a time shift, calculate:

$$\rho_1: 0\ 0\ 1\ 0\ 0\ 1\ 0\ 1\ 0\ 0\ 0\ 1\ 0\ 1\ 1\ 0\ 0$$
$$\rho_2: 1\ 0\ 0\ 1\ 0\ 1\ 0\ 0\ 0\ 1\ 0\ 1\ 1\ 0\ 0\ 0\ 0$$

- **5.4.1** Their distance using the spike time metric, $d_q^{time}(\cdot)$ for $q = 0$ and $q = \infty$.
- **5.4.2** Their distance using the interspike interval metric, $d_q^{inter}(\cdot)$ for $q = 0$ and $q = \infty$. Compare the results with the ones obtained for $d_q^{time}(\cdot)$.

5.5. Taking into account the following spike trains:

$$\rho_1: 0\ 1\ 0\ 0\ 1\ 0\ 1\ 0\ 0\ 1\ 0\ 0\ 1\ 0\ 1\ 0\ 0$$

ρ_2: 1 0 1 0 0 1 0 0 1 0 1 1 0 1 0 0 1 0

5.5.1 *Find their minimum distance using the interspike interval metric, $d_q^{inter}(\cdot)$, for $q = \infty$. (Note that there is more than one way to count the coincident interspike intervals.)*

5.5.2 *Find their distance using the interspike interval metric, $d_q^{inter}(\cdot)$, for $q = 0$.*

5.5.3 *Find their distance using the spike time metric, $d_q^{time}(\cdot)$, for $q = 0$ and $q = \infty$. Compare the results with those obtained for $d_q^{inter}(\cdot)$.*

5.6. Using Matlab, implement Algorithm 5.1 for the spike time metric and evaluate the distances between the spike trains of Problem 5.4 and Exercise 5.5.

5.7. Implement Algorithm 5.2 in Matlab for the interspike interval metric and evaluate the distances between the spike trains of Problem 5.4 and Problem 5.5.

5.8. Repeat the procedure in Exercise 5.1, but now measure the distance between the spike trains using the spike train metrics.

5.9. Implement Algorithm 5.5 for the spike events metric. Apply the spike events metric to the spike train sets obtained from Exercise 5.1. Repeat with real spike trains.

5.10. After implementing Algorithm 5.5 for the spike events metric, repeat the procedure from Problem 5.1 but calculate the error with the spike events metric.

Chapter 6

Design and Implementation of Bioelectronic Vision Systems

The ultimate goal for researching computational models for the retina is to develop bioelectronic vision systems to aid patients suffering from blindness. In the last years, research projects and consortia have been setup by joining people from different fields, such as biomedical engineering, computer science and electrical engineering. This chapter provides a brief overview of the most important projects and prostheses, which are categorized according to the two main types of bioelectronic vision systems considered in this book (see Chap. 1): retinal prostheses, based on epiretinal and subretinal implants, and cortical visual prostheses. Moreover, it describes and discusses in more detail a prototype of a bioelectronic vision system designed in the scope of the *Cortical Visual Neuroprosthesis for the Blind* European project.

6.1 Retinal Prostheses

People with retinal degeneration diseases, such as retinitis pigmentosa or macular degeneration, lose their sight as the cells in the retina that normally sense light deteriorate. Retinal implants can take over for these lost cells, converting light into neural signals that are transmitted through the optic nerve and interpreted by the brain. Simple versions of these devices have already been tested in humans, giving patients the ability to detect light and to distinguish simple objects. The epiretinal approach has the advantage of leaving the retina intact by placing the implant in the vitreous cavity, a natural fluid-filled space, and stimulating the ganglion cell layer. On the contrary, the subretinal implantation of a retinal prosthesis only replaces potentially damaged photoreceptors with a microstimulator by taking advantage of the remaining healthy cells of the retina. This type of implant requires detaching the retina in a more complex surgery, which is a problem since it is a fragile tissue only 0.25 mm thick.

6.1.1 Epiretinal Implants

The epiretinal approach to the retinal prosthesis involves the capture and digitization of images with an external device, such as a video camera. These images are transformed into patterns of electrical signals used to excite remaining viable inner retinal neurons. Power delivery and data telemetry sub-systems are required to drive this process. Several research groups have designed epiretinal implants based on intraocular and external elements with different characteristics: *i*) the Intraocular Retinal Prosthesis (IRP), developed by Mark Humayun and colleagues at the University of Southern California [Humayun et al. (1996)]; *ii*) Joseph Rizzo and John Wyatt developed an epiretinal prosthesis at the Harvard Medical School and the Massachusetts Institute of Technology [Wyatt and Rizzo (1996)]; and *iii*) the Retina Implant (RI) was developed by Rolf Eckmiller with a consortium of 14 expert groups in Germany [Eckmiller (1997)].

The IRP uses an external camera to acquire the image, which is passed through a visual processing unit to generate the information that is coded in the form of electrical pulses patterns. These patterns are transmitted into the eye by an inductive link telemetry system, composed of magnetic coils implanted in the temporal skull. The electrical stimulation pattern is delivered, via a transscleral (across eye wall) cable, to an array of microelectrodes attached to the inner retinal surface to stimulate viable inner retina neurons (intraocular part of the prosthesis). The first array stimulates inner retinal neurons through 16 platinum microelectrodes, ranging in size from 250 μm to 500 μm. Recent technological advances allowed the number of microelectrodes to increase to 60 and to incorporate microelectromechanical systems (MEMS) in order to achieve a better fit of a planar electrode array onto a curved inner retinal surface.

Clinical trials testing chronic long-term implantation of the IRP have been performed in 6 patients implanted with 16-electrode arrays manufactured by the Second Sight Medical Products [Javaheri et al. (2006)]. Reports of these trials show that patients are able to distinguish the direction of motion; they also have the ability to discriminate between percepts created by different electrodes, and the retinotopic organization is not lost when a patient loses sight [Lakhanpal et al. (2003)].

The retina implant team (EPI-RET) has developed the RI and founded the Intelligent Implants company in 1998. The implant is composed of a Retina Encoder (RE), a wireless Signal-and-Energy transmission system (SE), and a Retina Stimulator (RS). The RE, which is located outside the eye, consists of a photosensor array with around 100,000 pixels at the input. The RE approximates the typical receptive field properties of retinal ganglion cells by means of hundreds to thousands of tunable spatiotemporal filters. This cell output is encoded and transmitted via a wireless, signal and energy transmission system (electromagnetic and/or optoelectronic) to the implanted RS. The RS is a ring-shaped, soft microcontact foil centered about the fovea that is fixed to the epiretinal surface to be in contact with a sufficient number of retinal ganglion cells to elicit electrical spikes. The RE also provides a perception-based interaction between the RE and the human subject in order to tune the various receptive fields' filter properties with information "expected" by the central visual system. Eckmiller and his group have been testing the RE/RS mainly in animals [Walter et al. (1999)]. They have chosen to focus their efforts on understanding the information processing requirements of both the retinal prosthesis and the brain in terms of a dialogue-based RE tuning [Eckmiller et al. (1999)]. Clinical trials have been primarily focused on testing the RI implant and dialogue-based RE tuning.

Rizzo and Wyatt epiretinal implants also consist of independent intraocular and extraocular units, without batteries implanted within the body and no wires penetrating the eye. The extraocular unit is composed of a tiny charged couple device (CCD) camera, a signal processing unit and a fixed-direction laser; all mounted on a pair of glasses. The output of the CCD camera/signal processing unit modulates the amplitude of the laser beam (820 nm wavelength). The extraocular unit runs with replaceable batteries to be kept in the patient's pocket. The intraocular unit consists of a photodiode array and a stimulator chip sandwiched around a flexible thick polyamide strip that supports the electrodes. The photodiodes are used to capture the processed signal from a laser pulse emitted from the glasses. The stimulator chip then delivers this information to the microelectrode array on the epiretinal surface of the eye.

Rizzo and Wyatt have applied implants in 5 blind patients with Retinitis Pigmentosa, and 1 normal-sighted patient who was scheduled for enucleation due to orbital cancer. Three different types of electrode arrays were tested, with different numbers of electrodes, size, and spacing of the peripheral electrodes. The reported results from short-term studies were not conclusive. By stimulating a single electrode above a threshold level, multiple phosphenes were often perceived by the blind subjects. However, by simultaneously stimulating multiple electrodes it was not possible to perceive even the simplest visual pattern, neither by blind nor by normal-sighted patients. Due to the problems that Rizzo and Wyatt found in epiretinal stimulation, they have abandoned the epiretinal approach and are now developing a subretinal approach, which is discussed in the next section.

6.1.2 Subretinal Implants

In the subretinal approach, a microphotodiode array has to be implanted between the bipolar cell layer and the retinal pigment epithelium. The main advantage of this type of implant is that the microphotodiodes of the subretinal prosthesis directly replace the functions of the damaged photoreceptor cells, while it is assumed that the retina's remaining intact neural network is still capable of processing the generated electrical signals.

Subretinal implants were proposed by: *i*) Alan and Vincent Chow, who have developed the Artificial Silicon Retina (ASR) microchip [Chow et al. (2004)]; *ii*) Eberhart Zrenner and a consortium of research universities using a MicroPhotoDiode Array (MPDA) [Zrenner et al. (1998)]; and, more recently, *iii*) Rizzo and Wyatt [Wyatt and Rizzo (2006)], who have developed a third type of subretinal prosthesis.

Alan and Vincent Chow of Optobionics Corp believed that a subretinal implant could function as a simple solar cell; therefore, their ASR Microchip was powered entirely by light entering the eye [Peyman et al. (1998)]. With a diameter of two millimeters, the ASR contains approximately 5000 microelectrode-tipped microphotodiodes used to convert incident light into electrical signals. These electrical impulses, in turn, stimulate any viable retinal neurons, which then process and send these signals to the visual processing centers in the brain via the optic nerve. Ophthalmologist Alan Chow and his team at Optobionics tried this approach in people in the year 2000. They implanted a silicon disk with 5000 microscopic solar cells, or photodiodes in one eye in 30 people. Most of these implant recipients have reported moderate to significant improvements in at least one aspect of visual function, including light sensitivity, size of visual field, visual acuity, or movement or color perception. Many doubts arise about these results, because the amount of current needed to actively stimulate retina ganglion cells is not in the range of the current obtained from a photodiode [Wickelgren (2006)]. In fact, Chow abandoned the concept that the ASR Microchip is effective as a prosthetic device, and now he suggests that the insufficient levels of current delivered from the implant may have a therapeutic as well as neuroprotective effect on otherwise dying retinal photoreceptors. Therefore, this device is best classified as a therapeutic device, rather than as a true retinal prosthesis.

The SubRet consortium has designed and fabricated various types of ultrathin and flexible MPDA devices, as well as CMOS-based chips with different pixel sizes and electrode configurations. The first generation of developed MPDAs, similar to the original work of Chow, consists of 20×20 μm^2 pixels on a 3 mm diameter crystalline silicon chip. After this first generation, this team developed a special deposition technique to produce very thin and flexible MPDAs that fit to the curvature of the eyeball. They used amorphous hydrogenated silicon which has a light absorption 20-30 times higher than crystalline silicon. Biomedical experiments conducted by Zrenner and his team made clear that a purely photovoltaic operation is not effective, so additional energy has to be provided by near infrared or radio frequency power transmission. A charge transfer of 100-1000 $\mu C/cm^2$ within 1 ms is required for provoking a retina response, while light exposure at a retina location does not exceed 0.1 $\mu W/cm^2$.

Prototypes of their subretinal devices, with external power source to supply energy for the subretinal implant, have been proposed, namely using near infrared radiation. By implanting their prosthesis in rabbits, cats, and pigs, they attempted to detect electrically stimulated activity in the visual cortex as a result of retinal stimulation, as well as investigate the long-term biocompatibility and stability of these implants in the subretinal space. In nearly half of the tested animals, cortical evoked potentials were recorded with chronically implanted epidural electrodes during stimulation with light flashes, as well as during electrical stimulation of the subretinal space.

As stated in the last section, Rizzo and Wyatt decided to start working on subretinal implants. Although the Boston Retinal Implant Project is in the early stages of development, they have reported biocompatibility studies as well as the evaluation of surgical methods to implant their device in rabbits, pigs, and dogs. Minimally invasive surgical techniques have been tested, by using a posterior, ab externo approach to implant the prosthesis and to insert the stimulating electrode array in the subretinal space.

The main advantage of subretinal implants in comparison to the epiretinal approach is that microphotodiodes directly replace the functions of the damaged photoreceptor cells, while the retina's remaining intact neural network is still capable of processing electrical signals. However, the closer proximity of the subretinal prosthesis to inner retinal neurons predisposes the contacted retinal neurons to an increased likelihood of thermal injury resulting from heat dissipation. Together with the lack of external sources of energy for the microphotodiodes, this is one of the main problems of invasive neuroprostheses based on subretinal implants.

Open issues are the long-term biocompatibility of microelectronics in the saline environment of the eye, both in terms of hermetic packaging of the microfabricated electrode arrays and the heat generated and dissipated with its use. Also included in these biocompatibility issues is the unknown effect of chronic electrical stimulation on the retina.

6.2 Retinal Bioelectronic Vision System Design

Some numbers about visual acuity that can be useful for designing bioelectronic vision systems are: normal visual acuity (20/20) corresponds to angular separation of lines about 1 min of arc or spatial separation on the retina of about 10 μm; applying the Nyquist sampling frequency, for such visual acuity the maximum pixel size is 5 μm. Sufficient acuity (20/100) for reading with some visual aid requires pixels smaller than 25 μm. To achieve a useful reading performance, it has been estimated that about 600 pixels is the minimum for resolving images in the central field [Margalit et al. (2002)].

The design and implementation of bioelectronic vision systems for epiretinal and subretinal prostheses have been a topic of research during the last few years. A number of issues have to be addressed in order to design and implement bioelectronic vision systems based on retinal neuroprostheses. One important aspect is the interface between the electrode array and the retina, namely regarding biocompatibility and the requirement to conform with the spherical, concave surface of the retina. It is also necessary to supply power to permanent implants through a wireless system, since no wires are expected to go through the eye wall. Moreover, the electrical stimulation's pulse rate and the instant of occurrence need to be determined in a general way, but may need to be individually tuned for each patient.

Since it is not in the scope of this book to discuss in detail the design and implementation of the circuits and the systems, we choose to give an overview of the architectures of an intraocular epiretinal prosthesis test device [Scribner et al. (2001)], and of a proposed system that can be used

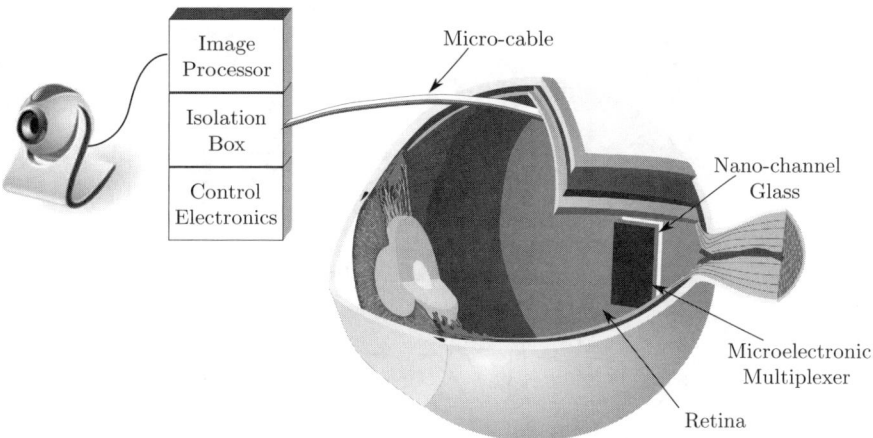

Fig. 6.1 Intraocular test device.

for both epiretinal and subretinal stimulation [Loudin et al. (2007)]. This latter system is based on a photodiode array implant, and video frames are processed and conveyed onto the retinal implant by a head-mounted near-to-eye projection system operating at near-infrared wavelengths.

The intraocular test device will enable short-term (less than an hour) human experiments to study issues related with interfacing electrode arrays with retinal tissue. The design combines two technologies: *i*) electrode arrays fabricated from NanoChannel Glass (NCG), and *ii*) Infrared Focal Plane Array (IRFPA) multiplexers. Figure 6.1 shows an IRP test device to be used in acute human experiments. Ophthalmologists use standard retinal surgical techniques in an operating room environment to perform the experimental procedure. Local anesthesia is administered so that the patient is conscious during the procedure.

NCG uses fiber optic fabrication techniques to produce thin wafers of glass with millions of very small channels with a diameter on the order of 1 μm, perpendicular to the plane of the wafer. These channels are filled with a good electrical conductor, and one surface of the glass is shaped with a spherical form consistent with the radius of the retina curvature. The image is serially input into the multiplexer via a very narrow flexible microcable. The real function of the micro-electronic multiplexer in Fig. 6.1 is essentially the reverse of the IRFPAs microelectronic multiplexers, acting as a demultiplexer to read an image onto the stimulator array. The electrical connection to the silicon multiplexer is made so there is nothing protruding

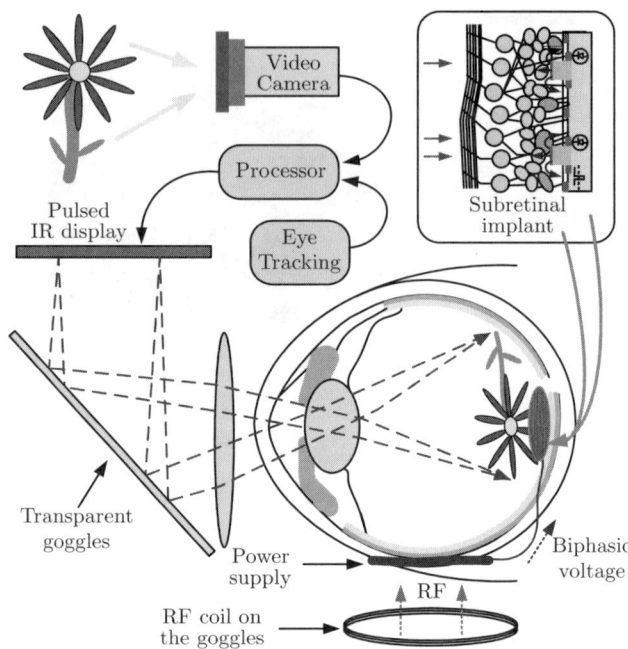

Fig. 6.2 A simplified layout of a retinal implant (adapted from [Loudin et al. (2007)]).

above the spherical curved envelope defined by the polished NCG surface, and therefore protects the retina from damage.

Because the test duration with the IRP experimental device is very short, there was no need to address the more difficult chronic issues that arise with permanent implants. Since patients are connected to external instrumentation during experiments, electrical shocks are prevented by isolating them using low voltage batteries and optocouplers.

A simplified layout of the general system architecture for implementing both epiretinal and subretinal protheses is presented in Fig. 6.2. It includes a goggles-mounted video camera, an image processor and a near-infrared (NIR) display. An extraocular power supply is connected to the subretinal implant. The inset shows a magnified view of a small area of the retinal implant.

The proposed video camera acquires and transmits 640×480 pixel images at 25-50 Hz to a pocket PC. This computer processes data and displays the resulting video on an liquid crystal display (LCD) matrix mounted on goggles worn by the patient. The LCD screen is illuminated with pulsed NIR (NIR wavelength: 800-900 nm) light, projecting each video image through the eye optics onto the retina. The NIR light is received by the photodiode array on an implanted chip, where each photodiode converts the NIR signal into a proportional electric current which is injected into the retina through an electrode placed in its center.

The projected NIR image is superimposed onto a normal image of the scene observed through the transparent goggles. Therefore, electrical stimulation introduces visual information into the retinal tissue above the implant, while any remaining peripheral vision responds normally to visible light. Such overlay is possible because NIR light does not activate normal photoreceptors, and the implant's response to natural visible light in the eye is negligible when compared to the bright and pulsed infrared image.

The prosthesis provides stimulation with a frame rate of up to 50 Hz in a central $10°$ visual field, with a full $30°$ field accessible via eye movements. Pixel sizes are scalable from 100 μm to 25 μm, which allows an acuity up to 20/100 to be achieved, which corresponds to 640–10,000 pixels on an implant with 3 mm in diameter.

Charge injection is maximized by biasing the photodiodes using a common pulsed biphasic power supply. Since the stimulation pulse must be synchronized with the IR light pulse, the system requires both power delivery and a trigger signal. Delivering 20 μA, 0.5 ms pulses to 640 electrodes at 50 Hz requires a peak current of about two tenths of a milliampere. The power transmission system is composed of a pair of inductively coupled coils: the transmitter coil is mounted beside the eye on the goggles, while the receiving coil and associated electronic circuit are implanted on the eye. The operating frequency of this transmission system is limited to 1 MHz. The AC current from the receiving coil is rectified using a half-wave rectifier, which collects charge into a tantalum electrolytic capacitor to provide DC current to the rest of the circuit.

One of the most important characteristics of this bioelectronic vision system is the fact that it can be used both for epiretinal and subretinal stimulation. The optical projection of the images into the eye also preserves a natural link between eye movements and visual information. Given that video goggles project images onto a retinal area much larger than the chip itself, a larger field of view can be observed with natural eye movements. Moreover, the parallel optical transmission of information during stimulation avoids the use of multiple wires connecting the acquisition system to the electrode array. The main disadvantage of this system is the fact that the photodiodes are placed in series with the electrodes, which prevents the generation of some types of typical biphasic stimulation waveforms, such as the symmetric biphasic current pulses.

6.3 Cortical Visual Prostheses

The first attempt to demonstrate the feasibility of a multichannel cortical neuroprosthesis by exploiting the retinotopic organization of the visual cortex was undertaken by Brindley and Lewin in 1968. They permanently implanted an array of 80 platinum disc electrodes subdurally onto the pial surface of the visual cortex of a volunteer who had been blind for more than a year. It was concluded that spatially discernable phosphenes could be evoked in 32 electrodes, the location of a phosphene roughly corresponded to the position of the stimulating electrode, and that small sets of phosphenes could be evoked by concurrently stimulating few electrodes.

In the last decades of the twentieth century, several researchers have studied and analyzed the neurophysiological principles that allow the production of phosphenes. One of these important studies was performed by Daniel Pollen [Pollen (1975)] by analyzing phosphenes production when transcranial magnetic stimulation was applied. In those years, several blind volunteer subjects were permanently implanted with cortical surface electrode arrays, namely by Dobelle [Dobelle (2000)].

Before attempting to construct a permanent implant, Dobelle decided to conduct a series of acute experiments involving volunteers undergoing other neurosurgeries [Dobelle (1974)]. Phosphene positions were mapped, thresholds were determined, and different stimulus parameters were tried. Some attempts were also made to combine single phosphenes into crude visual patterns. Two blind volunteers were implanted in 1978 at the Columbia-Presbyterian Medical Center in New York City, and they have both retained their implants for more than 20 years.

A platinum foil ground plant is perforated with an array of 5 mm diameter holes with 1 mm centers flat platinum electrodes centered in each hole, as depicted in Fig. 6.3. The ground plane eliminates phosphene interactions when multiple electrodes are stimulated simultaneously, and provides an additional measure of electrical safety that is not possible when stimulating between cortical electrodes and a ground plane outside the skull. Each electrode is connected by a separate teflon insulated wire to a connector contained in a carbon percutaneous pedestal. Stimulation delivered to each electrode typically consists of a train of six pulses delivered at 30 Hz to pro-

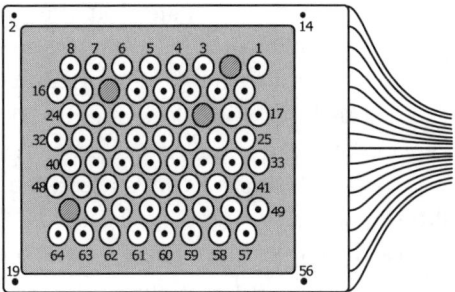

Fig. 6.3 Layout of the array of electrodes (looking through the electrodes).

duce the stimulation corresponding to an image frame. Frames have been produced with 1-50 pulses, with frame rates varying from 1 to 20 frames-per-second (fps). From experience, typical values are 4 fps, which includes trains containing only a single pulse. Biphasic symmetric 1 ms pulses are applied with threshold amplitudes between 10-20 V, which may vary up to $\pm 20\%$ from day to day; the system was calibrated on a daily basis.

It has been shown that these cortical implants allow blind people to recognize patterns when several phosphenes are induced in parallel. Over time, they were even able to perform simple tasks, such as recognize and find objects with different forms in simple scenarios. However, to stimulate

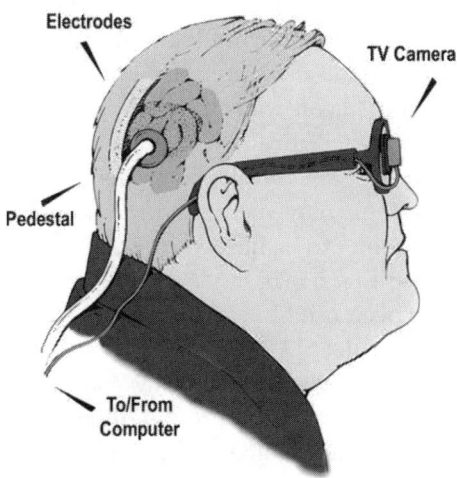

Fig. 6.4 A diagram of the Dobelle apparatus layout of a cortical implant.

this type of surface electrode, which is placed at the surface of the visual cortex, the current has to be sufficiently high (on the order of mA) in order to induce an electrical field able to stimulate internal neurons of the visual cortex. Moreover, the distance between electrodes restricts the resolutions, preventing the perception of more complex patterns. Another main disadvantage of the surface cortical visual prothesis described by Dobelle is that the apparatus includes a connector supported on a pedestal implanted in the skull, which receives the cables from the processing device and convey the signals to the implant in the cortex. This connector can be a source of infections, which can lead to serious health problems.

Intracortical microstimulation is a more recent technique used for developing cortical visual prostheses. This technique is based on the insertion of microelectrodes in the visual cortex, with 1 mm to 2 mm long. The insertion allows the deep layer of neurons to be directly stimulated, decreasing the required current several orders of magnitude, from mA to μA. These arrays of microelectrodes not only considerably reduce the required current, but also increase safety and reduce the distance between electrodes; this allows the increase of spatial resolution of the phosphene patterns. The laboratories that have developed the most relevant research work in intracortical implants are from the Illinois Institute of Technology [Troyk et al. (2006)], the National Institutes of Health [Schmidt et al. (1996)], and the University of Utah [Normann et al. (1999)].

At the University of Utah, the team of Richard Normann has designed a microelectrode array for recording and stimulating single cells in the cortex. It consists of 100 microelectrodes with 400 μm spacing, each 1.5 mm long and fabricated on a silicon wafer that measures 4 mm on a side. An electron micrograph of this silicon-based microelectrode array is presented in Fig. 6.5(a). A technique was developed to insert the electrodes into the cerebral cortex with a single movement, and when inserted, it allows a high density of stimulation points to be achieved and it stimulates up to layer IV of the visual cortex (see Fig. 6.5(b)).

This array has also been used in the Cortical Visual Neuroprosthesis for the Blind (CORTIVIS) project [Ahnelt et al. (2002)] referred to in Chap. 1. The main objective of this project was to show the feasibility of an artificial vision system capable of conferring to profoundly blind people some kind of vision, namely the discrimination of shape and location of objects, resulting in a substantial improvement in the standard of living of blind and visually impaired persons.

A scheme of a bioelectronic vision system, with the main modules of the

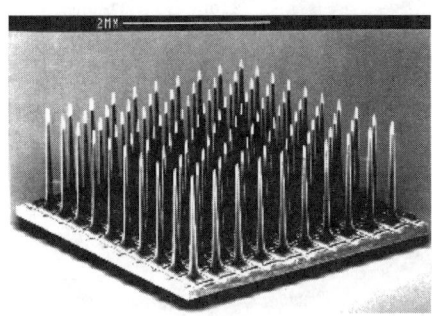
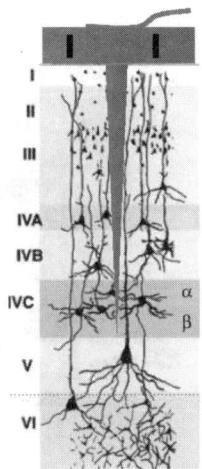

(a) Microphotograph of the microelectrodes.

(b) Inserted microelectrode.

Fig. 6.5 The Utah Microelectrode Array.

CORTIVIS cortical neuroprosthesis, is represented in Fig. 6.6. This system includes an image capture device, often a video camera, that captures the visual stimulus and converts it into electrical signals, usually electrical currents or voltages. These electrical signals are sampled, quantized and processed in the *Neuromorphic Encoder* module. This module is responsible for all the required digital processing, including the generation of the spike events used to stimulate the visual cortex neurons. The information about these events is serialized and can be transmitted to inside the skull through electrical connectors, as in [Dobelle (2000)], or through a wireless communication system. In both cases, the *Electrode Stimulator* module responsible for exciting the visual cortex cells through a microelectrode array has to be implanted in the visual cortex by neurosurgery. However, while the latter approach reduces the risk of infections, and therefore improves the patient health level, it poses the need for wireless delivery of both power and data [Piedade *et al.* (2005)].

In the next sections, we describe and analyze in more detail the bioelectronic vision system developed to implement an intracortical visual neuroprosthesis in the scope of the CORTIVIS project.

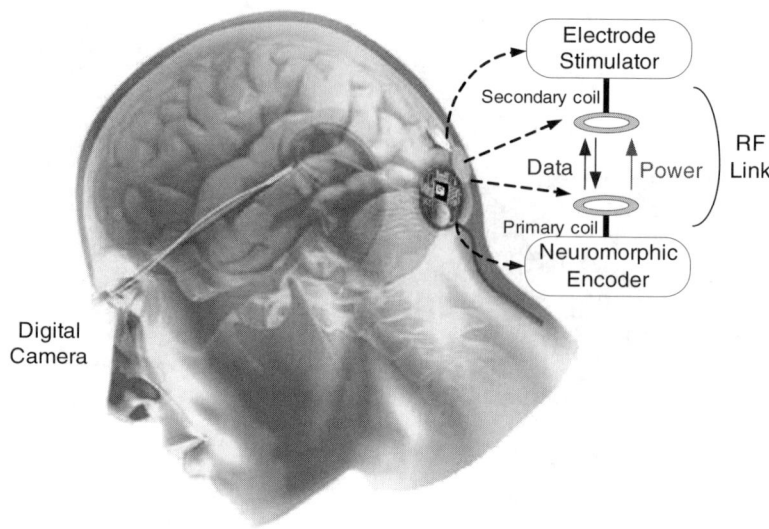

Fig. 6.6 Bioelectronic vision system based on the CORTIVIS neuroprosthesis (from [Piedade et al. (2005)]).

6.4 Cortical Bioelectronic Vision System Design

The prototype of the visual neuroprosthesis herein presented uses a digital video camera, a visual encoder module based on a functional retina model, namely the deterministic model described in Chap. 4, and a microelectrode stimulator. A serial link with a dedicated wireless communication system delivers both power and data into the human head. As shown in Fig. 6.7, three main sub-modules can be identified in the *Bio-inspired Visual Encoder* module: *i*) the *Early Layers* of the retina model that applies spatial and temporal filtering to the input signals, in order to compute the firing rate corresponding to the observed light patterns; *ii*) the *Neu-*

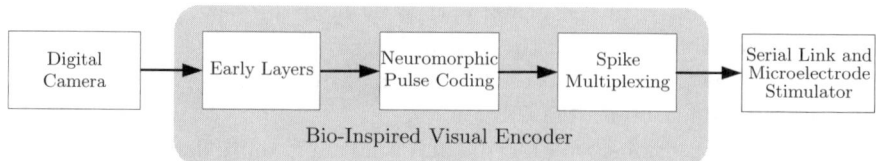

Fig. 6.7 Modules of the bio-inspired processing module of the artificial retina.

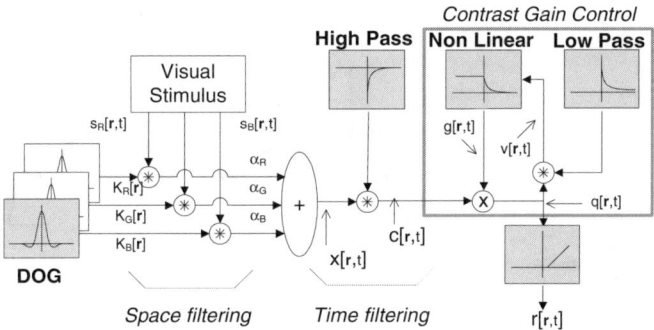

Fig. 6.8 Extended RGB model for replacing natural human retina processing.

romorphic Pulse Coding devoted to convert the output of the retina *Early Layers* to a series of spike events; and *iii*) a final sub-module that performs *Spike Multiplexing* to put together the information, corresponding to the spike events generated at different spacial locations, in a serial stream. The *Serial link* is established to transmit the information to the *Microelectrode Stimulator* implanted in the visual cortex. First, the visual encoder module and its main components are described, and subsequently the characteristics of the remaining modules are presented.

6.4.1 Early Layers

In Fig. 6.8, the visual stimulus $s(\mathbf{r}, t)$ is represented by the basic color components (**R**ed, **G**reen and **B**lue) of the pixels acquired by a digital camera; when just the intensity of the stimulus is considered, $s(\mathbf{r}, t) = s_R(\mathbf{r}, t) = s_G(\mathbf{r}, t) = s_B(\mathbf{r}, t)$. To reproduce the response of the receptive fields stimulated in the different color channels, the spatial filter (DoG) in Fig. 4.6 can be expanded into three different spatial filters dedicated to each of the color channels $\{R, G, B\}$, as depicted in Fig. 6.8. Equation (4.40) is computed by assigning individual values to the parameters of each channel.

A parallel architecture for implementing the model in Fig. 6.8 requires a large amount of hardware resources. The total number of multipliers required for 2D DoG filtering a pixel is equal to the number of color channels times the size of the convolution matrix – an impulse response with $N \times N$ non-null coefficients requires N^2 multipliers. For example, for $N = 7$, the number of multipliers is 49 for grayscale images and 147 when each

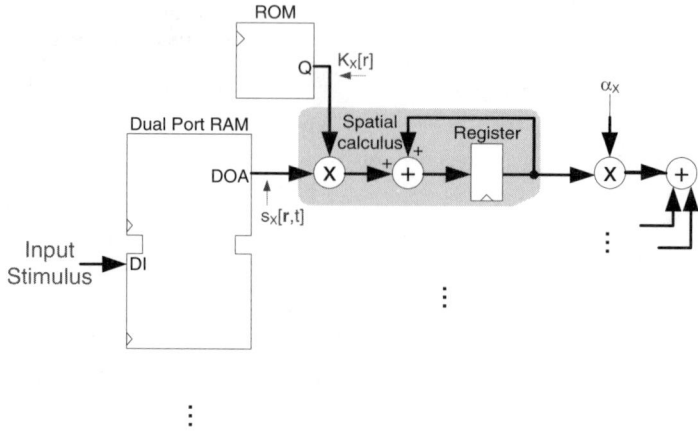

Fig. 6.9 Architecture for computing the spatial filter.

RGB channel is individually considered. The folding technique can be applied to reduce the number of multipliers, and in the limit, a single multiplier can be used per color channel. In this case, three multipliers for the R,G, and B channels. This serial architecture adopted in the CORTIVIS project requires $N \times N$ cycles to convolve the input samples with the DoG coefficients; therefore, the effective frequency of such circuit is N^2 times slower than its operating frequency.

Figure 6.9 represents the architecture for calculating a receptive field, corresponding to one of the color channels. In this figure, the dual port random access memory (RAM) is used to asynchronously transfer the current stimuli from the camera and read the previously stored stimuli. The read only memory (ROM) stores the receptive field's filter coefficients, which correspond to the DoG coefficients. Since the number of target microelectrodes is usually lower than the number of pixels in the acquired image, a decimation process is also applied at this stage.

All address and control signals required by the digital circuit in Fig. 6.9 are generated by the control circuit, which implements a pre-defined state machine. The calculus of the spatial part of the model will then be completed by adding the results from all three receptive fields.

The signal flow graph (SFG) for the remaining components of the *Early Layers* model are represented in Fig. 6.10. The SFG adopted for the temporal high pass filter allows a reduction in the amount of required memory. This is a very important aspect, since a time delay (z^{-1}) corresponds to a

Design and Implementation of Bioelectronic Vision Systems 215

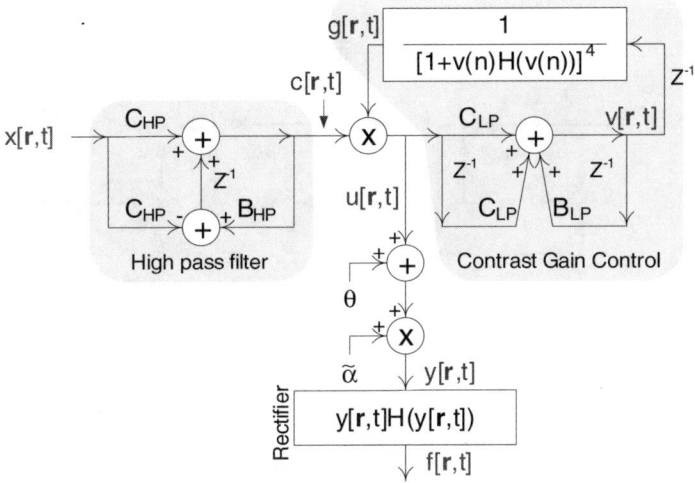

(a) Data diagram 1.

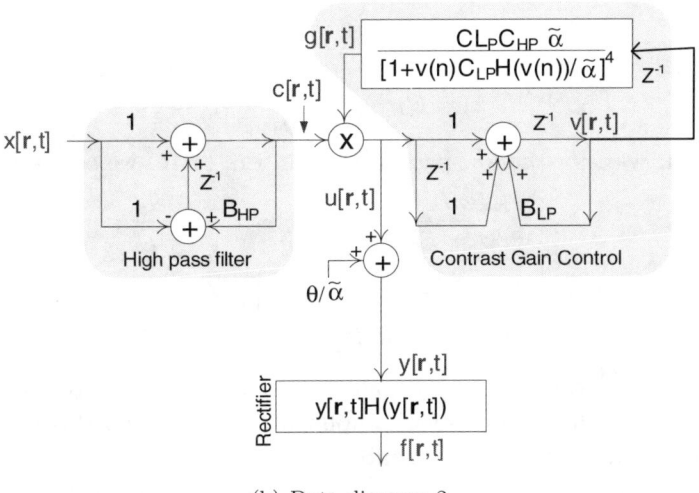

(b) Data diagram 2.

Fig. 6.10 Signal flow graphs for the temporal (high pass filter and contrast gain control) and rectifier components for the *Early Layers* model.

full frame. If in Eq. (4.43) we consider

$$C_{HP} = \frac{2}{2+\alpha T_s} \quad \text{and} \quad B_{HP} = \frac{2-\alpha T_s}{2+\alpha T_s},$$

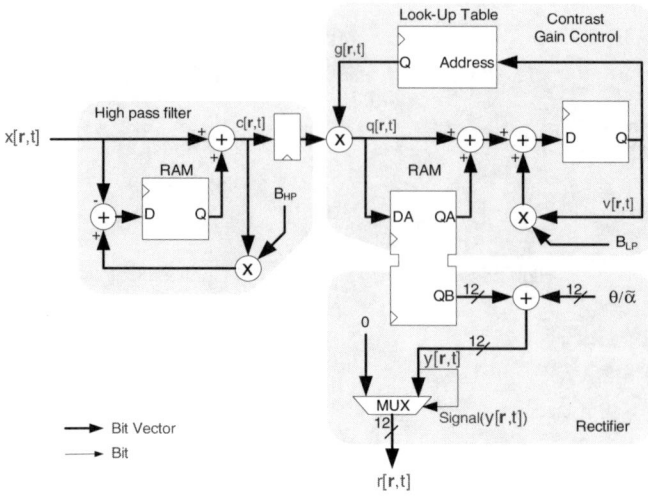

Fig. 6.11 Architecture of the temporal part of the *Early Layers* model.

we obtain

$$K_t(z) = C_{HP} \frac{1 - z^{-1}}{1 - B_{HP}z^{-1}} \quad . \tag{6.1}$$

In a similar way, for the low-pass filter, if in Eq. (4.46) we consider

$$C_{LP} = \frac{B\tau T_s}{2\tau + T_s} \quad \text{and} \quad B_{LP} = \frac{2\tau - T_s}{2\tau + T_s} \; ,$$

we obtain

$$H(z) = C_{LP} \frac{1 + z^{-1}}{1 - B_{LP}z^{-1}} \quad . \tag{6.2}$$

In addition, to save hardware resources, the C_{HP}, C_{LP} and $\tilde{\alpha}$ constants were grouped with the non-linear amplification block, and this function is implemented by a single look-up-table (Fig. 6.10(b)).

The development of the architecture for the temporal part of the *Early Layers* module (Fig. 6.11) follows the same approach adopted for the spatial part, but now time delays are replaced by RAM blocks. The non-linear amplification is implemented through a look-up-table, while the rectifier is implemented through a multiplexer, which selects the signal $y[\mathbf{r}, t]$ when it is above zero and 0 otherwise.

By analyzing the cost of the components for implementing the *Early Layers* model, it can be concluded that the most expensive one is the memory required by the time filters; each delay in these filters corresponds to a

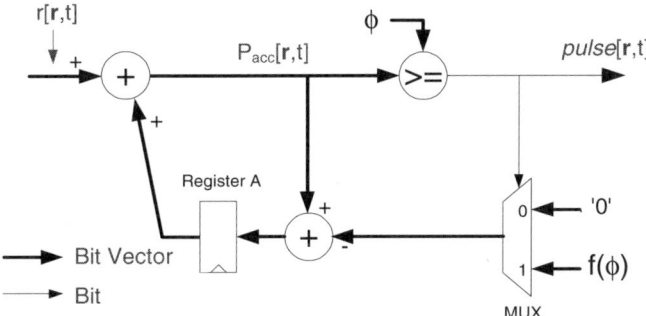

Fig. 6.12 Simplified architecture of the *Neuromorphic Pulse Coding* module; one register per microelectrode.

full frame. In our case, the amount of memory required to implement the time filters is equal to the number of delays times the number of bits per sample times the number of target microelectrodes.

6.4.2 *Neuromorphic Pulse Coding*

The *Neuromorphic Pulse Coding* module of the *Visual Encoder* system is responsible for the generation of spikes, from the output firing rate generated in the *Early Layers*. Figure 6.12 shows an architecture for generating a spike train based on the method described in Sec. 3.4.2. The architecture in this figure, for a simple model where the width of the pulses is constant, uses register A as an accumulator and the multiplexer to implement the Heaviside functions.

Assuming that $r[\mathbf{r}, t] = r$ is constant during the integration period, the number of cycles, N, necessary for a fire event is

$$r \times N = \phi. \tag{6.3}$$

However, N is also the ratio between the spike event frequency, f_{fire}, and the clock frequency, f_{CLK}:

$$\phi = r \times \frac{f_{CLK}}{f_{fire}}. \tag{6.4}$$

Equation (6.4) results in the fire event frequency equalling the output of the *Early Layers*, r, when

$$\phi = f_{CLK}. \tag{6.5}$$

To code pulses for multiple cells, the number of registers in Fig. 6.12 has to be multiplied by the number of microelectrodes, forming RAM blocks, such as in the *Early Layers*. Once again, a single processing block is used and shared in time since the usage of multiple processing blocks implies a significantly increase in the amount of required hardware. Therefore, the frequency will be raised by the number of microelectrodes (M) and Eq. (6.5) will be changed to

$$\phi = f_{sample} = \frac{f_{CLK}}{M} \; . \tag{6.6}$$

By assuming the restriction of a maximum spike firing of 1 kHz, it is necessary for the module clock frequency to be at least

$$\{f_{CLK}\}_{min} = \{f_{sample}\}_{min} \times M = 1 \times M \text{ kHz} \; . \tag{6.7}$$

6.4.3 Spike Multiplexing

The *Spike Multiplexing* block represented in Fig. 6.7 is responsible for "collecting" the spike events from the different cells and to multiplex them in time into a single stream. The AER [Boahen (2000)] is designed to be a neuromorphic representation of signals, by simulating the biological axons which can multiplex a large number of spike events in a single digital bus. The AER protocol adopted in the CORTIVIS project defines a way to asynchronously send information from one set of cells to another set of cells without time stamps. When one of these cells wants to trigger a spike event in a receiver neuron, it would request the bus and put the appropriate address into the bus. At the receiving end, the cells were listening for when their addresses arrive on the bus and reacted accordingly with that event. There is no master clock on such an architecture to represent time, a spike is placed onto the bus as quickly as possible in order to minimize the latency. This feature goes along with the idea that precise spike timing is important: no spikes are lost due to collisions, and if multiple spikes happen simultaneously, they are transmitted as soon as possible in sequence.

An extended architecture for the AER protocol is represented in Fig. 6.13. The AER tree (Fig. 6.13(a)) consists of multiple stages of arbiters (Fig. 6.13(b)) which multiplex the spike requests from their inputs to their outputs. This circuit operates in sequence from the top *Neuromorphic Pulse Coding* (NPC) blocks that are responsible for generating requests to send spikes through the serial link to the corresponding microelectrodes. Each arbiter receives one or more spike requests, and through a generic decision method, places that request on the input of the decider of the next stage. The requests will then travel from stage to stage down

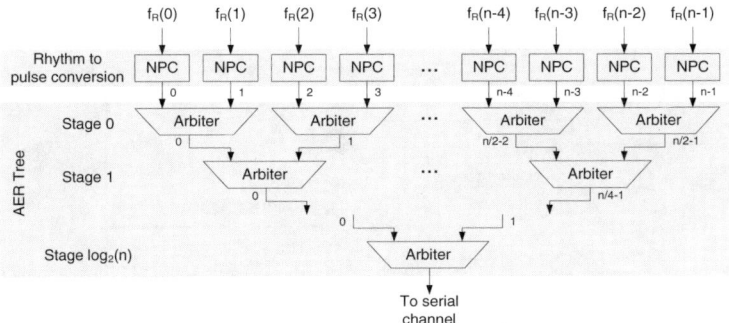

(a) Diagram of a full AER tree with 2 inputs and one output per arbiter (NPC-*Neuromorphic Pulse Coding* cell).

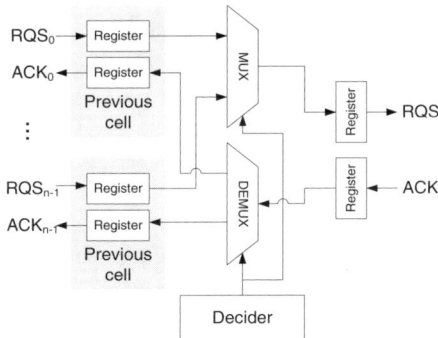

(b) AER arbiter with n inputs and one output and registers between stages.

Fig. 6.13 Structure of an AER tree and possible implementations for the arbiters.

to the last decider, which sends them through the communication channel. This type of architecture guarantees that no spike is overlapped by another, and the spikes will be sent as soon as possible to the serial channel.

However, a large amount of hardware is required to implement the architecture depicted in Fig. 6.13. The challenge in this module is to deliver pulses to the proper microelectrodes as soon as possible, according to the pulse generation sequence. Since in the earlier modules the spikes are not simultaneously produced - the computing resources are shared amongst all cells - the only thing needed to be saved is the spike information (such as the microelectrode address, and if necessary, the width and height). Therefore it is possible to create a First In First Out (FIFO) pipeline of spike events to replace the highly cost tree structure in Fig. 6.13.

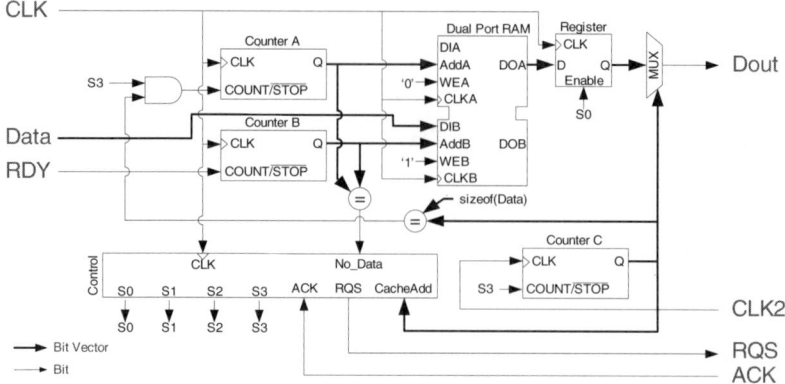

Fig. 6.14 Architecture of the FIFO based AER module.

Figure 6.14 shows the circuit corresponding to the alternative AER architecture that requires much less hardware. Information related to an event is processed through the following phases, every time one spike event is generated by the *Neuromorphic Pulse Coding* module (RDY is high); the data (pulse width and height) are stored on the RAM and *Counter B*, which addresses port B of the Dual Port RAM, is incremented. The read cycle is controlled in the FIFO based AER module by a *control* block. Whenever there are no spikes stored in the RAM, the state machine remains in its initial state, and once a spike is generated by the *Neuromorphic Pulse Coding* module, it is stored in the RAM and the system detects it (*Counter A* output becomes different than *Counter B* output). Next, it sends a RQS signal to the channel controller. Once the channel responds with the ACK signal, both the AER module and the channel are ready to send/receive the information. Meanwhile, when in the initial state, the pulse information was stored in an output *Register*; therefore guaranteeing that the information is not overwritten by the write section. *Counter C* sequentializes the data bits at the link's rate as the clock signal is supplied by the channel (CLK2). Once all bits have been sent, *Counter A* will increment its value and a new cycle starts. This circuit can also implement a *time out* mechanism: once the RAM is filled with spikes (the channel bandwidth is temporarily short) the system starts to overwrite spikes.

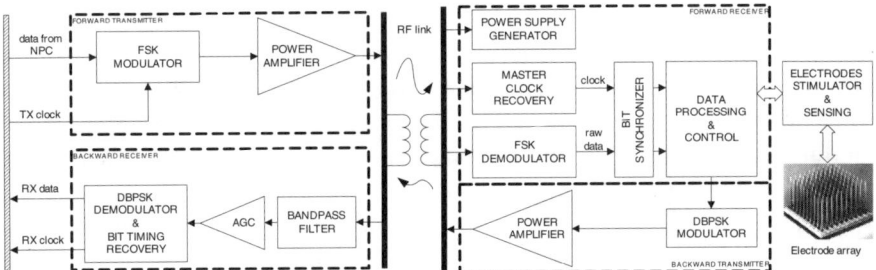

Fig. 6.15 Block diagram of the RF link and the *Microelectrode Stimulator* module (from [Piedade et al. (2005)]).

6.4.4 Serial Communication Link

Contrary to the Dobelle implant, in the CORTIVIS bioelectronic vision system there is no physical connection through the skull. The serial communication link represented in Fig. 6.6 is a wireless radio frequency (RF) link that provides both power and bidirectional data communication through a coupling transformer, with the primary unit located outside the body and a secondary unit implanted inside the skull. A block diagram of the circuits for implementing the RF link and the Microelectrode Stimulator module is presented in Fig. 6.15.

The power provided from the outside primary unit has to be enough to achieve the tenths of milliwatt of power required by the implant inside the skull. The provided power must be significantly higher than that required by the implant due to the inevitable coupling losses. The forward data link is established to send data to the implant, while a reverse data link is implemented for monitoring purposes. Different modulation schemes have been proposed for the transmission through the RF link, namely Amplitude Shift Keying (ASK), Frequency Shift Keying (FSK) and Binary Phase Shift Keying (BPSK). After a careful analysis of the different modulation characteristics, FSK was chosen as the modulation to use in the forward link, and BPSK was adopted for the reverse data link [Piedade et al. (2005)]. The most important reasons for this decision were the facts that the FSK receiver and the BPSK transmitter, which are located on the implant side, have the simplest electronic circuits and require the lowest power consumption. This decision also takes into account the fact that the performance of the ASK receiver is highly dependent on the amplitude of the received signal, which is unknown and depends on the relative position between the implant and the outside unit.

The *Forward Transmitter* receives the multiplexed spike data from the spike multiplexing block in a synchronous serial bit stream format at a rate of 1 Mbps. Access to the RF link is arbitrated in the FIFO buffer used to store the spike events until the link is able to send them. The FSK modulation is applied to data from a central frequency of 10 MHz and with a frequency deviation of $\Delta f = \pm 323$ kHz. This signal feeds a Class-E switching-mode tuned power amplifier, which was chosen to optimize the efficiency at the transmitter, as shown in Fig. 6.16(a).

The BPSK modulator uses a carry signal with 5 MHz with low amplitude (1 V) for transmitting at a bit rate of 156.25 kHz. This lower bit rate allows power consumption reduction and is enough to accomplish implant monitoring procedures, such as electrode impedance measurement and calibration. Details about the standard circuit topologies used to implement the modulator and demodulator, on the send and receiver sides, respectively, can be found in [Piedade *et al.* (2005)].

The coupling transformer is a key component of the RF link, which has to allow proper system operation for intercoil distances within certain limits, typically up to 2 cm. The real transformer exhibits a distributed parameter behavior, as depicted in the model presented in Fig. 6.16(b). The absence of an iron core makes it impossible to have a strong magnetic coupling, resulting in a high magnetic flux not connected with the secondary coil. In order to compensate for the equivalent inductances relative to the primary and secondary magnetic flux dispersions, capacitors C1 and C2 are connected in series with the respective coils, resonating at the carrier frequency. The RF transformer behaves like a double tuned bandpass filter, with a measured coupling factor of 0.3 for an intercoil distance of 1 cm, using carefully designed planar coils of Litz wire with a diameter about 3 cm.

The power supply generator referred to in Fig. 6.15 is a power recovery circuit composed of a half-wave rectifier, protection circuits and a series regulator. It recovers the required power from the received signal, with an efficiency of around 30% for an intercoil distance about 1 cm. After demodulation, the bit stream is fed to the *Bit Synchronizer* that provides a synchronized clock and data to the *Data Processing and Control* unit. The *Data Processing and Control* unit performs bit and frame synchronization and frame disassembly. Finally, the formatted data are forwarded to the *Microelectrode Stimulator* and *Sensing Block*.

The *Microelectrode Stimulator* module (see Fig. 6.6 and Fig. 6.15), which also has sensing functions, is composed of a set of digital-to-analog converters (DACs) [Santos *et al.* (2006)], which can be used to stimulate

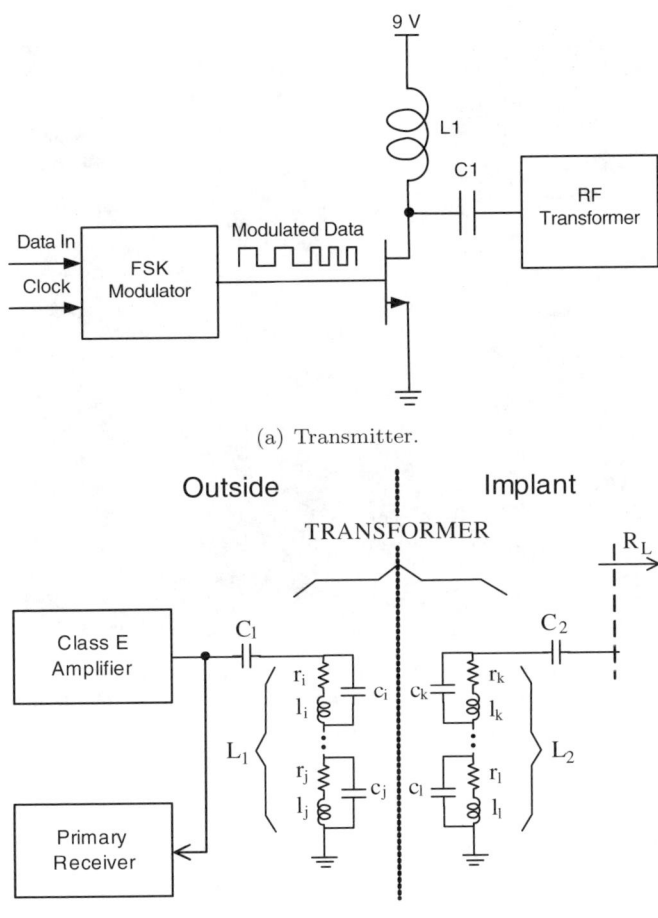

Fig. 6.16 The RF link (from [Piedade *et al.* (2005)]).

up to 1024 microelectrodes in the visual cortex. The amplitude and duration of the spikes are pre-registered in dedicated registers located on this module.

6.5 Vision Prosthesis Prototype

A complete prototype of a vision prosthesis was implemented and integrated in a demonstrator based on a motorized human model head named Elonica,

Fig. 6.17 a) Decimated original image; b) Spatial filtered image; c) Temporal filtered image; d) Recovered image from spikes; e) Processing Module; f) Elonica prototype system.

as depicted in Fig. 6.17.

The RF link in this prototype is a ten times down scaled-frequency version, operating with a forward data rate of 100 kbps over a carrier frequency of 1 MHz. This prototype was built using conventional integrated circuits and discrete components, and supports a backward data rate of 15.625 kbps with a carrier frequency of 500 kHz. Experimental results show that for a primary RF unit, with power consumption of 180 mW, it is possible to recover 50 mW of power at the secondary receiver output, which leads to an efficiency of about 28% for intercoil distance up to 1 cm. A neuromorphic encoder was also developed using an Field Programable Gate Array (FPGA). Although the visual neuroprosthesis system was planned to support driving up to 1024 electrodes, the scaled-frequency of the prototype supports the stimulation of about 100 microelectrodes.

Pictures in Fig. 6.17(a)-(d) correspond to photographs from images displayed on a VGA monitor. Figure 6.17(f) shows the CMOS standard XVGA (1024×768 pixel) digital color microcamera placed behind Elonica's

Table 6.1 FPGA circuit area (from [Piedade et al. (2005)]).

Modules	Slice Occupation		RAM blocks	Mult.
Total available resources	3584	100%	16	16
Image Capture and Resize	108	3%	0	2
Acquisition and image resize	142	4%	0	0
Retina Model	420	12%	5	3
Spike multiplexing	60	2%	2	0
Image Display	218	6%	8	9
Complete System	948	26%	15	12

glasses, which is connected to the retina processor by a flat cable digital bus placed within a tissue strap. The designed artificial retina processing board in Fig. 6.17(e) is based on a XILINX SPARTAN XC3S400 FPGA [Specification (2003)]. The circuitry implemented in the FPGA includes the control of the microcamera, a hardware module to decimate the input image and a processor module that generates three additional processed images. The following hardware modules were implemented: *i)* low pass filtering and decimation of the original color image (Fig. 6.17a) integrated in the image capture module; *ii)* the spatial filtering module that implements the spatial component of the processing in the *Early Layers* (Fig. 6.17b); *iii)* retinal temporal filtering of the images required to compute the firing rate (Fig. 6.17c); and *iv)* the module for Neuromorphic Encoder and Spike Multiplexing. Figure 6.17d) presents the recovered black and white image obtained by decoding the generated spikes, which requires simulating the operation of the human brain. These four images are composed in the Image Display module that generates a full frame VGA compatible image that is fed to a standard LCD monitor.

The artificial retina blocks were described in VHDL using the XILINX WEBPACK 6.2 tool. The resource occupancy of the FPGA for the implemented modules is shown in Table 6.1. It can be seen that even in this modest FPGA, the complete system was synthetized using only 25% of the total resources, operating at a maximum frequency of 85 MHz. This greatly exceeds the required 1.5 MHz to implement the Early Layers block (to process a 32×32 image by executing 49 operations per cycle due to hardware folding), and about 1 MHz for the Neuromorphic Pulse Coding block (spikes generated at a maximum rate of 100 Hz).

A prototype of a bioelectronic vision system with similar features has been recently developed by Sawan and his team [Coulombe et al. (2007)].

Table 6.2 Main features of implants for visual neuroprostheses.

Prosthesis	Max n. sites	Modulation Forward/Back.	f_{carry}/f_{data} ($\times 10^6$)	Max pulses/s	Real-time Controller
CORTIVIS	1024	FSK/DBPSK	10/1	100k-1M	Hard
Sawan	992	OOK/LSK	13.56/1.5	500k	Hard/Soft

It is a scalable prototype with stimulation modules that generate biphasic current pulses supplied to a matrix of 16 electrodes. The small size stimulation modules are connected to an interface module, located away from the stimulation sites. Table 6.2 summarizes the main characteristics of the implants dedicated to visual prostheses developed on the CORTIVIS project and by Sawan and his team. Alternatively, On-Off Keying (OOK) and Load-Shift Keying (LSK) modulation schemes are used for the forward and backward communication, respectively.

Although the two prototypes show some differences, it can be observed that the maximum number of stimulation sites, the carrier and data frequencies, and the pulse rate assume similar values, which indicates that although this technological area is recent, it has already achieved a certain degree of maturity.

6.6 Conclusions and Further Reading

This last chapter of the book focused on bioelectronic vision systems and results from the contribution of several different areas of the sciences and engineering, such as biology, neuroscience, signal processing and microelectronics. These bioelectronic vision systems can be supported in retinal implants or in cortical implants.

Retinal implants have advantages over cortical implants with regard to surgical implantation and access to target nerve cells. The mapping of the retina, which lies in the back of the eye, to a physical location in space is well known. The types of retinal implants are delineated primarily by the anatomical location of the electrode neuron interface, which can be on the epiretinal surface or in the subretinal space. In this last type of implant, the microphotodiodes directly replace the functions of the damaged photoreceptor cells, while the retina's remaining intact neural network processes the electrical signals. Epiretinal implants rely on external imaging devices, including a camera to acquire an image and an external image processor.

Regarding epiretinal prostheses, subretinal prostheses do not require external cameras or image processing units, and the patient's eye movements can be used to locate objects. However, the lack of an external source of energy for the microphotodiodes in the subretinal prosthesis is a significant drawback, since the levels of ambient light are not sufficient for the current generated by the microphotodiodes to stimulate the adjacent retinal neurons.

Clinical trials testing chronic longterm implants have been performed since the early 2000s in humans, both for subretinal and epiretinal prostheses [Javaheri et al. (2006)]. In the trial periods, which vary from a few months to several years, it was demonstrated that patients perceived phosphenes in response to the electrical stimulation of the retina. Biocompatibility studies were performed in order to examine the effects of an extraneous body in the subretinal space, and surgical methods to implant the device have been evaluated as well. It has been demonstrated that the idea behind the simple subretinal approach is not effective because it lacks a viable source of power [Zrenner (2002)]. It is believed that the low levels of current delivered from the implant, although insufficient to electrically activate any remaining retinal neurons in a retina with damaged photoreceptors, may have therapeutic and neuroprotective effects for otherwise dying retinal photoreceptors.

Experiments have been reported in animals with prototypes of subretinal devices that contain external power sources. This power source supplies energy to the subretinal implant by means of very thin wires that run outside of the eye [Sachs et al. (2005)]. Systems have been also designed to be used for both epiretinal and subretinal stimulation. In one of these systems [Loudin et al. (2007)], near infrared light is sent to a photodiode array, which converts it into an electrical current that is injected into the retina. Charge injection is increased by biasing the diodes using a common power supply. In this system, the camera is mounted in the goggles, and the power transmission system consists of a pair of inductively coupled coils; the transmitter coil is mounted beside the eye on the goggles, while the receiving coil and associated electronic circuit are implanted in the eye.

For retinal blindness with degeneration of the ganglion cell neuron, which in turn gives rise to the optic nerve axons, a retinal prosthesis would not be helpful. Therefore, visual cortical prostheses have been pursued by a number of individual researchers and groups since the 1960s; it was experimentally shown that phosphenes could be evoked by stimulating electrodes implanted in the visual cortex.

Cortical prostheses are much more complex, not only because they have to account for the significant signal processing that must be implemented, but also due to the challenge of positioning electrodes precisely in the primary visual cortex. Dobelle was the first to develop a cortical neuroprosthesis and to implant blind volunteers with permanent electrode arrays in the visual cortex. This cortical neuroprosthesis includes a television camera, which is mounted into a pair of sunglasses, linked to a sub-notebook computer in a belt pack. The belt pack also contains another microcontroller and associated electronics to stimulate the brain. This stimulus generator is connected through a percutaneous pedestal to the planar electrodes on the visual cortex, in this case an array of about 64 surface electrodes. Although it has been shown that the cortical implants allow blind people to recognize patterns when several phosphenes are induced in parallel, significant drawbacks are associated with this approach to developing visual prostheses. Some of the difficulties include interactions between phosphenes and multiple phosphenes induced by a single electrode, as well as the usage of high currents and large electrodes. Occasionally, pain was caused by meningeal stimulation, and possible focal epileptic activity was induced by electrical stimulation [Pollen (1977)]. Another main disadvantage of the surface cortical visual prosthesis described by Dobelle is that the apparatus includes a connector implanted in the skull, which can lead to health problems due to infections.

A second generation of intracortical prostheses has been developed in order to overcome the main drawbacks identified in the first generation of intracortical visual prostheses. This type of visual neuroprosthesis performs intracortical microstimulation through an array of microelectrodes implanted into the primary visual cortex. The space between microelectrodes is quite reduced and the microelectrodes are inserted in depth to directly stimulate the neurons in the inner layers of the visual cortex. Microelectrode arrays are fabricated on a silicon wafer, and typically have a spacing of hundreds of micrometers and depths of about 1 mm [Maynard et al. (1997)]. This new generation of intracortical prostheses makes use of wireless implants to multichannel microstimulation [Sawan et al. (2005)]. Wireless inductive links carry power and stimulus information to inside the cranium, usually by means of a low-coupling transformer. This wireless link establishes a path between the primary unit, located outside the body, and the secondary unit, implanted inside the body. In this chapter, we have presented in detail the intracortical prosthesis developed in the scope of the *Cortical Visual Neuroprosthesis for the Blind* (CORTIVIS) project. In this prosthesis, the primary unit comprises a neuromorphic encoder, a forward transmitter and a backward receiver. The developed neuromorphic encoder generates the spikes to stimulate the cortex by mimicking the characteristics of the spatiotemporal receptive field response of ganglion cells. The secondary unit comprises a forward receiver, a microelectrode stimulation circuitry and a backward transmitter that is used to monitor the implant. A prototype of the proposed system was developed and tested in animals. However, clinical trials have to be further performed in order to test the implantation of the microelectrode array and the operation of an entire intracortical prosthesis.

All of this research has paved the way toward restoring useful vision to profoundly blind people by interfacing a cortical neuroprosthesis with the visual cortex. However, several issues have to be addressed by the research groups, such as the biocompatibility of microelectronics, the heat generated and dissipated by the electronic devices and the plasticity of the visual system in response to electrical stimulation. Attention also has to be given to understanding how the brain interprets a stimulation pattern resulting from an increasing number of electrodes, which is also a crucial issue in the evolution of vision prosthetic design.

Last, but not least, it is important to have the involvement of companies to produce, manufacture and test medical devices. Once such devices are available, the research community will be able to perform visual psychophysical experiments, in order to develop stimulation algorithms that result in the "best" perceptions.

Exercises

6.1. *Discuss the main advantages and disadvantages of subretinal implants in comparison to epiretinal implants.*

6.2. *Figure 6.18 presents a Matlab source code for displaying an image sequence stored in an AVI file and copying it to another AVI file. Prepare the Matlab execution environment in your computer.*

> **6.2.1** *Find an AVI file, execute the code in Fig. 6.18 and observe the results.*
>
> **6.2.2** *Use the Matlab function* **text** *to add a caption to the movie. Suggestion: use the Matlab* **text** *function, e.g.,*
> text(info.Width/2,info.Height/2,'ORIGINAL MOVIE','color',[1 1 1],'HorizontalAlignment','center')

```
input_file = 'InputFile.avi'; output_file = 'OutputFile.avi';

% Getting the properties the AVIFILE
info = aviinfo(input_file);

% Creating an AVIFILE for writing
aviobj = avifile(output_file);

% Setting the number of frames per second of the output AVIFILE
aviobj.fps = info.FramesPerSecond;

% Create figure to display results and set its properties
fig=figure;
set(fig,'DoubleBuffer','on','Position',[1 1 info.Width info.Height]);
set(gca,'NextPlot','replace','Visible','off','Position',[0 0 1 1]);

% iterate for all frames
for frame_num=1:info.NumFrames
    % Update figure title
    str=sprintf('\nFRAME: %d (%.2f sec.) of %d (%d sec.)',...
        frame_num, frame_num/info.FramesPerSecond,info.NumFrames,...
        round(info.NumFrames/info.FramesPerSecond));
    set(fig,'name',str); axis off;

    % Reading a frame from avifile
    frame = aviread(input_file,frame_num);

    % Showing the frame
    image(frame.cdata);

    % Adding a frame to the output AVIFILE
    outframe = getframe;
    aviobj = addframe(aviobj,outframe);
end

% Closing output file
aviobj = close(aviobj);
```

Fig. 6.18   Matlab code to read and write an AVI file.

6.2.3 *Display four tiled images with different captions. Suggestion: present the images on a 2× 2 array organization.*

6.3. *Implement the model in Fig. 6.8 in Matlab step by step according to the following script:*

6.3.1 *Program a function to perform spatial filtering of a frame according to Eq. (4.40). Default values: $\sigma_C/\Delta r = 0.8$, and $\sigma_S/\Delta r = 1.6$.*

6.3.2 *Program a function to perform temporal filtering given the current and previous spatial filtered frames according to Eq. (4.44). Default value: $\alpha = 4$ ($1/T_s = $ frame rate).*

6.3.3 *Program the contrast gain control (CGC) block in Fig. 6.8, and apply it to the temporal filtered frames. Default value: $\tau = 0.17$.*

6.3.4 *Implement the model by instantiating the programmed functions in the proper order.*

6.4. *Generate a movie displaying four images, the original and the three images generated by the functions programmed in Exercise 6.3, by using the function implemented in Exercise 6.2.*

6.5. *Write a function in Matlab for implementing the Neuromorphic Pulse Coding (NPC) – see Fig. 6.12 – in order to generate the spike trains from the firing rate for each microelectrode (consider that each pixel in a frame corresponds to a microelectrode for this and the following exercises).*

6.6. *Applying the NPC function to the spatially processed frame:*

6.6.1 *Generate a movie displaying the original frame, the spatially processed frame and the produced sequence of spikes.*

6.6.2 *Recover the original spatially processed frame by applying a low pass filter (suggestion: apply a very narrow band low-pass filter).*

6.7. *Derive an expression to obtain the memory requirements in function of the frame size to implement the model in Fig. 6.8.*

6.8. *Estimate the bandwidth required for transmitting a sequence of spikes, given the maximum firing rate and the number of pixels. Consider that a fixed number of bits is used to represent a spike.*

6.9. *Consider the development of a visual neuroprostheses.*

6.9.1 *Enumerate the main drawbacks of using wired communication to transmit information from the neuromorphic encoder to the implant in the cortex.*

6.9.2 *Present the main differences between using surface electrodes (Fig. 6.3) and implanted micro-electrodes (fig 6.5)*

6.9.3 *Discuss the main disadvantage of using Amplitude Shift Keying (ASK) modulation for transmitting data in the RF link of 6.6.*

6.10. *Considering the AER implementation presented in Fig. 6.11, derive an expression for the time required for the FIFO memory to overflow as a function of the following parameters: i) frame size; ii) firing rate (assume a constant firing rate for all pixels); iii) channel bit rate. Suggestion: the rate at which spike events are removed from the FIFO is equal to the channel bit rate divided by $\lceil \log_2(\text{frame size}) \rceil$.*

Bibliography

Abramowitz, M. and Stegun, I. A. (1965). *Handbook of Mathematical Functions* (Dover Publications, New York, USA).

Agnew, W. and McCreery, D. (eds.) (1990). *Neural Prostheses: Fundamental Studies*, Biophysics and Bioengineering Series (Prentice Hall, Englewood Cliffs, New Jersey, USA).

Ahnelt, P., Ammermüller, J., Pelayo, F., Bongard, M., Palomar, D., Piedade, M., Ferrandez, J., Borg-Graham, L. and Fernandez, E. (2002). Neuroscientific basis for the design and development of a bioinspired visual processing front-end, in *IFMBE Proceedings*, Vol. 3 (Vienna), pp. 1692–1693.

Arfken, G. B. and Weber, H. J. (2005). *Mathematical Methods for Physicists*, 6th edn. (Academic Press, San Diego, USA).

Arno, P., Capelle, C., Wanet-Defalque, M.-C., Catalan-Ahumada, M. and Veraart, C. (1999). Auditory coding of visual patterns for the blind, *Perception* **28**, 8, pp. 1013–1029.

Bair, W., Koch, C., Newsome, W. and Britten, K. (1994). Power spectrum analysis of bursting cells in area MT in the behaving monkey, *The Journal of Neuroscience* **14**, 5, pp. 2870–2892.

Berry, M. J., Warland, D. K. and Meister, M. (1997a). The structure and precision of retinal spike trains, *Neurobiology, Proc. Natl. Acad. Sci. USA* **94**, pp. 5411–5416.

Berry, M. J., Warland, D. K. and Meister, M. (1997b). The structure and precision of retinal spike trains, *Neurobiology* **94**, pp. 5411–5416.

Boahen, K. A. (2000). Point-to-point connectivity between neuromorphic chips using address-events, *IEEE Transactions on Circuits and Systems* **47**, 5, pp. 416–434.

Brélen, M. E., Duret, F., Gérard, B., Delbeke, J. and Veraart, C. (2005). Creating a meaningful visual perception in blind volunteers by optic nerve stimulation, *IOP Publishing, Journal of Neural Engineering* **2**, pp. S22–S28.

Brenner, N., Agam, O., Bialek, W. and de Ruyter van Steveninck, R. (2002). Statistical properties of spike trains: Universal and stimulus-dependent aspects, *Physical Review* **66**.

Brindley, G. S. and Lewin, W. S. (1968). The sensations produced by electrical stimulation of the visual cortex, *The Journal of Physiology* **196**, pp. 479–493.

Chichilnisky, E. J. (2001). A simple white-noise analysis of neuronal light responses, *Network: Computation in Neural Systems* **12**, 2, pp. 199–213.

Chow, A. Y., Chow, V. Y., Packo, K. H., Pollack, J. S., Peyman, G. A. and Schuchard, R. (2004). The artificial silicon retina microchip for the treatment of vision loss from retinitis pigmentosa, *Archives of Ophthalmology* **122**, 4, pp. 460–469, URL http://archopht.ama-assn.org/cgi/content/abstract/122/4/460.

Coulombe, J., Sawan, M. and Gervais, J. (2007). A highly flexible system for microstimulation of the visual cortex: Design and implementation, *IEEE Transactions on Biomedical Circuits and Systems* **1**, 4, pp. 258–269.

Curcio, C. A., Sloan, K. R., Kalina, R. E. and Hendrickson, A. E. (1990). Human photoreceptor topography, *The Journal of Comparative Neurology* **4**, 292, pp. 497–523.

Cyberkinetics (2008). Neurotechnology systems, inc, http://www.cyberkineticsinc.com.

Dayan, P. and Abbot, L. F. (2001). *Theoretical Neuroscience: Computational and Mathematical Modeling of Neural Systems* (The MIT Press, Cambridge, Massachusetts, USA).

de Ruyter van Steveninck, R. and Bialek, W. (1988). Real-time performance of a movement-sensitive neuron in the blowfly visual system: Coding and information transfer in short spikes sequences, *Proceedings of the Royal Society of London, Series B, Biological Sciences* **234**, 1277, pp. 379–414.

Dobelle, W. (1974). Phosphenes produced by electrical stimulation of human occipital cortex, and their application to the development of a prosthesis for the blind, *The Journal of Physiology* **243**, 2, pp. 553–576.

Dobelle, W. (2006). Artificial vision for the blind, http://www.seeingwithsound.com/etumble.htm.

Dobelle, W. and Mladejovsky, W. (1974). Phosphenes produced by electrical stimulation of human occipital cortex, and their application to the development of a prosthesis for the blind, *Journal of Physiology* **243**, pp. 553–576.

Dobelle, W. H. (2000). Artificial vision for the blind by connecting a television camera to the visual cortex, *American Society for Artificial Internal Organs (ASAIO) Journal* **46**, pp. 3–9.

Dong, D. W. and Atick, J. J. (1995). Statistics of natural time-varying images, *Network: Computation in Neural Systems* **6**, 3, pp. 345–358.

Donoghue, J. P. (2002). Connecting cortex to machines: recent advances in brain interfaces, *Nature Neuroscience* **5**, pp. 1085–1088.

Dowling, J. E. (1987). *The retina: an approachable part of the brain* (Belknap Press).

Eckmiller, R. (1997). Learning retina implants with epiretinal contacts, *Ophtalmic Research* **29**, 5, pp. 281–289.

Eckmiller, R., Hunermann, R. and Becker, M. (1999). Exploration of a dialog-based tunable retina encoder for retina implants, *Neurocomputing* **26–27**, pp. 1005–1011.

Efstratiadis, A. and Koutsoyiannis, D. (2002). An evolutionary annealing-simplex algorithm for global optimisation of water resource systems, in I. Publishing (ed.), *Hydroinformatics 2002: Proceedings of the Fifth International Conference on Hydroinformatics* (Cardiff, United Kingdom), pp. 1423–1428.

Eggermont, J. J. (1998). Is there a neural code? *Neuroscience and Biobehavioral Reviews* **22**, 2, pp. 355–370.

Flannery, B. P., Teukolsky, S. A., Press, W. H. and Vetterling, W. T. (2002). *Numerical Recipes in C*, 2nd edn. (Cambridge University Press, Cambridge, Massachusetts, USA), the Art of Scientific Computing.

Gerstner, W. and Kistler, W. (2002). *Spiking Neuron Models: Single Neurons, Populations, Plasticity* (Cambridge University Press).

Gestri, G., Mastebroek, H. A. K. and Zaagman, W. H. (1980). Stochastic constancy, variability and adaptation of spike generation: Performance of a giant neuron in the visual system of the fly, *Biological Cybernetics (Springer-Verlag)* **38**, 1, pp. 31–40.

Grill-Spector, K. and Malach, R. (2004). The human visual cortex, *Annual Review of Neuroscience* **27**, pp. 649–677.

Grumeta, A. E., Jr., J. L. W. and Rizzo III, J. F. (2000). Multi-electrode stimulation and recording in the isolated retina, *Journal of Neuroscience Methods, Elsevier* **101**, pp. 31–42.

Hayes, M. H. (1996). *Statistical Digital Signal Processing and Modeling*, 4th edn. (John Wiley& Sons, New York, USA).

Hessburg, P. and Rizzo III, J. (2007). The eye and the chip. World congress on artificial vision 2006, *Journal of Neural Engineering* **4**, 1, pp. 1–2.

Humayun, M., Jr, E. J., Dagnelie, G., Greenberg, R., Propst, R. and Phillips, D. (1996). Visual perception elicited by electrical stimulation of retina in blind humans, *Archives of Ophthmology* **114**, pp. 40–46.

Humayun, M. S., de Juan Jr., E., Weiland, J. D., Dagnelie, G., Katona, S., Greenberg, R. and Suzuki, S. (1999). Pattern electrical stimulation of the human retina, *Vision Research* **39**, 15, pp. 2569–2576.

Ifeachor, E. C. and Jervis, B. W. (2002). *Digital Signal Processing: A Practical Approach*, 2nd edn. (Prentice Hall, Harlow, England).

Javaheri, M., Hahn, D., Lakhanpal, R., Weiland, J. and Humayun, M. (2006). Retinal prostheses for the blind, *Annals Academy of Medicine Singapore* **35**, 3, pp. 137–144.

Keat, J., Reinagel, P., Reid, R. C. and Meister, M. (2001). Predicting every spike: A model for the responses of visual neurons, *Neuron* **30**, pp. 803–817.

Kolb, H. (2003). How the retina works, *Scientific American* **91**, pp. 28–35.

Kolb, H., Fernández, E. and Nelson, R. (2002). WebVision: The organization of the retina and visual system, http://webvision.med.utah.edu.

Lakhanpal, R., Yanai, D., Weiland, J., Fujii, G., Caffey, S., Greenberg, R. R., Jr, J. E. and Humayun, M. (2003). Advances in the development of visual prostheses, *Current Opinion in Ophthalmology* **14**, pp. 122–127.

Lim, J. S. (1990). *Two-Dimensional Signal and Image Processing* (Prentice-Hall, Englewood Cliffs, New Jersey, USA).

Loudin, J., Simanovskii, D., Vijayraghavan, K., Sramek, C., Butterwick, A., Huie,

P., McLean, G. and Palanker, D. (2007). Optoelectronic retinal prosthesis: System design and performance, *Journal of Neural Engineering* **4**, pp. S72–S84.

Margalit, E., Maia, M., Weiland, J. D., Greenberg, R. J., Fujii, G. Y., Torres, G., Piyathaisere, D. V., O"Hearn, T. M., Liu, W., Lazzi, G., Dagnelie, G., Scribner, D. A., de Juan Jr, E. and Humayun, M. S. (2002). Retinal prosthesis for the blind, *Survey of Ophthalmology* **47**, 4, pp. 335–356.

Marmarelis, V. Z. (2004). *Nonlinear Dynamic Modeling of Physiological Systems* (Wiley-IEEE Press).

Martins, J. C. and Sousa, L. A. (2005). Comparison of computational retina models, in *5th IASTED Conference on Visualization, Imaging, and Image Processing*, (Benidorm, Spain), pp. 156–161.

Martins, S. F., Sousa, L. A. and Martins, J. C. (2007). Additive logistic regression applied to retina modelling, *IEEE International Conference on Image Processing, 2007. ICIP2007*. **3**, pp. 309–312.

Maynard, E. M. (2001). Visual prosthesis, *Annual Review in Biomedical Engineering* **3**, pp. 145–168.

Maynard, E. M., Nordhausen, C. T. and Normann, R. A. (1997). The Utah intracortical electrode array: a recording structure for potential brain-computer interfaces. *Electroencephalography and Clinical Neurophysiology* **102**, 3, pp. 228–239.

Berry II, M. J., Brivanlou, I. H., Jordan, T. A. and Meister, M. (1999). Anticipation of moving stimuli by the retina, *Nature* **398**, pp. 334–338.

Berry II, M. J. and Meister, M. (1998). Refractoriness and neural precision, *The Journal of Neuroscience* **18**, 6, pp. 2200–2221.

Optobionics Corporation (2006). Optobionics: Technology for vision, http://www.optobionics.com.

Project CORTIVIS (2006). Cortical visual neuroprosthesis for the blind, http://cortivis.umh.es/.

Rizzo III, J. F. and Wyatt, J. (1997). Prospects for a visual prosthesis, *The Neuroscientist* **3**, 4, pp. 251–262.

Vision Egg (2007). Visionegg, http://www.visionegg.org.

Meister, M. (2007). The Meister Lab, http://rhino.harvard.edu.

Meister, M. and Berry II, M. J. (1999). The neural code of the retina, *Neuron* **22**, pp. 253–450.

Meister, M., Pine, J. and Baylor, D. A. (1994). Multi-neuronal signals from the retina: acquisition and analysis, *Journal of Neuroscience Methods, Elsevier* **51**, pp. 95–106.

Mood, A. M., Graybill, F. A. and Boes, D. C. (1974). *Introduction to the Theory of Statistics*, 3rd edn. (McGraw-Hill, Singapore).

Moon, T. K. and Stirling, W. C. (2000). *Mathematical Methods and Algorithms for Signal Processing* (Prentice Hall, Upper Sadle River, New Jersey, USA).

Nirenberg, S., Carcieri, S. M., Jacobs, A. L. and Latham, P. E. (2001). Retinal ganglion cells act largely as independent encoders, *Nature* **411**, pp. 698–701.

Nirenberg, S. and Latham, P. E. (2003). Decoding neuronal spike trains: How important are correlations? *Proceedings of the National Academy of Sciences*

of the United States of America (PNAS) **100**, 12, pp. 7348–7353.
Normann, R. A., Maynard, E. M., Rousche, P. J. and Warren, D. J. (1999). A neural interface for a cortical vision prosthesis, *Vision Research* **39**, 15, pp. 2577–2587.
Oppenheim, A. V., Shafer, R. W. and Buck, J. R. (1999a). *Discrete-Time Signal Processing*, 2nd edn. (Prentice-Hall, Upper Saddle River, New York, USA).
Oppenheim, A. V., Willsky, A. S. and Nawab, S. H. (1999b). *Signals & Systems*, 2nd edn. (Prentice-Hall, Inc., Upper Saddle River, New Jersey, USA).
Orfanidis, S. J. (1990). *Optimum Signal Processing: An Introduction*, 2nd edn. (McGraw-Hill Book Co., Singapore).
Papoulis, A. and Pillai, S. U. (2002). *Probability, Random Variables, and Stochastic Processes*, 4th edn. (McGraw-Hill, New York, USA).
Peyman, G., Chow, A., Liang, C., Chow, V., Perlman, J. and Peachey, N. (1998). Subretinal semiconductor microphotodiode array, *Ophthalmic Surgery and Lasers* **29**, 3, pp. 234–241.
Pezaris, J. S. and Reid, R. C. (2007). Demonstration of artificial visual percepts generated through thalamic microstimulation, *Proceedings of the National Academy of Sciences of the United States of America (PNAS)* **104**, 18, pp. 7670–7675.
Piedade, M., Gerald, J., Sousa, L., Tavares, G. and Tomás, P. (2005). Visual neuroprosthesis: A non invasive system for stimulating the cortex, *IEEE Transactions on Circuits and Systems* , pp. 2648–2662.
Pillow, J. W., Paninski, L., Uzzell, V. J., Simoncelli, E. P. and Chichilnisky, E. J. (2005). Prediction and decoding of retinal ganglion cell responses with a probabilistic spiking model, *The Journal of Neuroscience* **25**, 47, pp. 11003–11013.
Pillow, J. W. and Simoncelli, E. P. (2003). Biases in white noise analysis due to non-Poisson spike generation, *Neurocomputing* **52–54**, pp. 109–115.
Pollen, D. (1975). *The Nervous System*, chap. Some perceptual effects of electrical stimulation of the visual cortex in man (Raven Press, New York, USA), pp. 519–528.
Pollen, D. (1977). Responses of single neurons to electrical stimulation of the surface of the visual cortex, *Brain, Behavior and Evolution* **14**, pp. 67–86.
Proakis, J. G. and Manolakis, D. K. (2006). *Digital Signal Processing* (Prentice-Hall, Englewood Cliffs, New Jersey, USA).
Purves, D., Augustine, G. J., Fitzpatrick, D., Hall, W. C., Lamantia, A.-S., McNamara, J. O. and Williams, S. M. (2007). *Neuroscience*, 4th edn. (Sinauer Associates, Inc., Sunderland, MA, USA).
Reich, D. S., Victor, J. D., Knight, B. W., Ozaki, T. and Kaplan, E. (1997). Response variability and timing precision of neuronal spike trains in vivo, *The Journal of Neurophysiology* **77**, 5, pp. 2836–2841.
Reinagel, P. (2001). How do visual neurons respond in the real world? *Current Opinion in Neurobiology* **11**, 4, pp. 437–442.
Rieke, F., Warland, D., de Ruyter van Steveninck, R. and Bialek, W. (1997). *Spikes: Exploring the Neural Code* (The MIT Press, Cambridge, Massachusetts, USA).

Rodieck, R. W. (1965). Quantitative analysis of cat retinal ganglion cell response to visual stimuli, *Vision Research, Pergamon Press* **5**, 12, pp. 583–601.

Rodieck, R. W. (1998). *The First Steps in Seeing* (Sinauer Associates; Sunderland, Massachusetts, USA).

Ross, S. M. (2006). *Simulation*, 4th edn. (Academic Press, San Diego, USA).

Rugh, W. J. (1981). *Nonlinear System Theory: The Volterra Wiener Approach* (Jonhs Hopkins Univ. Press., Baltimore, USA).

Rust, N. C., Schwartz, O., Movshon, J. A. and Simoncelli, E. (2004). Spike-triggered characterization of excitatory and suppressive stimulus dimensions in monkey V1 directionally selective neurons, *Neurocomputing* , pp. 793–799.

Sachs, H., Schanze, T., Brunner, U., Sailer, H. and Wiesenack, C. (2005). Transscleral implantation and neurophysiological testing of subretinal polyimide film electrodes in the domestic pig in visual prosthesis development, *Journal of Neural Engineering* **2**, S57–S64.

Santos, M., Fernandes, J. and Piedade, M. (2006). A microelectrode stimulation system for a cortical neuroprosthesis, in *Proc. of Conference Design of Circuits and Integrated Systems (DCIS'06)*.

Sawan, M., Hu, Y. and Coulombe, J. (2005). Wireless smart implants dedicated to multichannel monitoring and microstimulation, *IEEE Circuits and Systems Magazine* **5**, 1, pp. 21–39.

Schmidt, E., Bak, M., Hambrecht, F., Kufta, C., O'Rourke, D. and Vallabhanath, P. (1996). Feasibility of a visual prosthesis for the blind based on intracortical microstimulation of the visual cortex, *Brain* **119**, pp. 507–522.

Schwartz, O., Chichilnisky, E. J. and Simoncelli, E. P. (2002). Characterizing neural gain control using spike-triggered covariance, in T. G. Dietterich, S. Becker and Z. Ghahramani (eds.), *Advanced Neural Information Processing Systems (NIPS*01)*, Vol. 14 (MIT Press, Cambridge, Massachusetts, USA), pp. 269–276.

Scribner, D., Humayun, M., Justus, B., Merritt, C., Klein, R., Howard, J., Peckerar, M., Perkins, F., Johnson, L., Bassett, W., Skeath, P., Margalit, E., Eong, K.-G. A., Weiland, J., de Juan Jr., E., Finch, J., Graham, R., Trautfield, C. and Taylor, S. (2001). Intraocular retinal prosthesis test device, *23rd Annual International Conference of the IEEE Engineering in Medicine and Biology Society* .

Sellers, P. H. (1974). On the theory and computation of evolutionary distances, *SIAM Journal Applied Mathematics* **26**, 4, pp. 787–793.

Shoham, S. (2001). *Advances towards an implantable motor cortical interface*, Ph.D. thesis, University of Utah.

Simoncelli, E. P., Paninski, L., Pillow, J. and Schwartz, O. (2004). *Characterization of Neural Responses with Stochastic Stimuli*, chap. 23 (MIT Press), pp. 327–338.

Smith, S. W. (2003). *Digital Signal Processing: A Practical Guide for Engineers and Scientists* (Newnes, Elsevier Science, Burlignton, MA, USA).

Specification, A. P. (2003). *Spartan-3 FPGA Family: Complete Data Sheet*, XILINX, San Jose, CA, USA, ds099 edn.

Therrien, C. W. (1992). *Discrete Random Signals and Statistical Signal Processing* (Prentice-Hall, Englewood Cliffs, New Jersey, USA).
Thiel, A., Wilke, S. D., Greschner, M., Bongard, M., Ammermüller, J., Eurich, C. W. and Schwegler, H. (2003). *Temporally Faithful Representations of Salient Stimulus Movement Patterns in the Early Visual System*, chap. Visual Attention Mechanisms (Kluwer Academic Publishers), pp. 93–100.
Tomás, P. and Sousa, L. (2008). Statistical analysis of a spike train distance in Poisson models, *IEEE Signal Processing Letters* **15**, 1, pp. 357–360.
Troyk, P., Detlefsen, D. and DeMichele, G. (2006). A multifunctional neural electrode stimulation ASIC using neurotalk interface, in *Proceedings of the 28th IEEE EMBS Annual International Conference* (Canada), pp. 2994–2997.
Uzzell, V. J. and Chichilnisky, E. J. (2004). Precision of spike trains in primate retinal ganglion cells, *Journal of Neurophysiology, The American Physiological Society* **92**, pp. 780–789.
van Rossum, M. C. W. (2001). A Novel Spike Distance, *Neural Comp.* **13**, 4, pp. 751–763.
Victor, J. D. (1999). Temporal aspects of neural coding in the retina and lateral geniculate: a review, *IOP Publishing, Network: Comput. Neural Syst. 10 (1999)* **10**, pp. R1–R66.
Victor, J. D. (2005). Spike train metrics, *Current Opinion in Neurobiology* **15**, 5, pp. 585–592.
Victor, J. D. and Purpura, K. P. (1996). Nature and precision of temporal coding in visual cortex: A metric-space analysis, *Journal of Physiology* **76**, 2, pp. 1310–1326.
Victor, J. D. and Purpura, K. P. (1997). Metric-space analysis of spike trains: theory, algorithms and application, *Network: Computational Neural Systems* **8**, 2, pp. 127–164.
Walter, P., Szurman, P., Vobig, M., Berk, H., Ludtke-Handjery, H., Richter, H., Mittermayer, C., Heimann, K. and Sellhaus, B. (1999). Successful long-term implantation of electrically inactive epiretinal microelectrode arrays in rabbits, *Retina* **19**, 6, pp. 546–552.
Wandell, B. A. (1995). *Foundations of Vision* (Sinauer Associates, Sunderland, Massachusetts, USA).
Warren, D. J. and Normann, R. A. (2003). *Visual Neuroprosthesis* (CRC Press, Boca Raton, Florida, USA).
Weiland, J. D., Liu, W. and Humayun, M. S. (2005). Retinal prosthesis, *Annual Review Biomedical Engineering* **7**, pp. 361–401.
Weiss, P. (2001). The seeing tongue, *Science News* **160**, 9, p. 140.
Werblin, F. and Roska, B. (2007). The movies in our eyes, *Scientific American*, pp. 73–79.
Westwick, D. T. and Kearney, R. E. (2003). *Identification of Nonlinear Physiological Systems*, IEEE Press Series on Biomedical Engineering (IEEE Press, Piscataway, New Jersey, USA), ISBN 0-471-27456-9.
Wickelgren, I. (2006). A vision for the blind, *Science* **312**, pp. 1124–1126.
Wilke, S. D., Thiel, A., Eurich, C. W., Greschner, M., Bongard, M., Am-

mermüller, J. and Schwegler, H. (2001). Population coding of motion patterns in the early visual system, *Journal of Comparative Physiology A: Sensory, Neural, and Behavioral Physiology* **187**, 7, pp. 549–558.

World Health Organization (2004). Magnitude and causes of visual impairment, Tech. Rep. Fact Sheet n. 282, United Nations Organization.

Wu, N. J. (2006). Bionic eye, URL- http://www.svec.uh.edu/BIONIC.html.

Wulf, M. (2001). *On Modeling the Spatiotemporal Processing Characteristics of the Retina: What is the Retina for?*, Ph.D. thesis, Universität Hamburg.

Wyatt, J. and Rizzo, J. (1996). Ocular implants for the blind, *IEEE Spectrum*, pp. 47–53.

Wyatt, J. and Rizzo, J. (2006). Development of a wireless first generation boston retinal implant subretinal prosthesis, in *Proceedings of The Eye and the Chip - 2006 World Congress on Artificial Vision*.

Yu, H.-H. and de Sa, V. R. (2004). Nonlinear reverse correlation with synthesized naturalistic noise, *Neurocomputing, Computational Neuroscience: Trends in Research 2004* **58–60**, pp. 909–913.

Ziemer, R. E., Tranter, W. H. and Fannin, D. R. (1998). *Signals and Systems: Continuous and Discrete*, 4th edn. (Prentice-Hall, New York, USA).

Zrenner, E. (2002). Will retinal implants restore vision? *Science* **295**, 5557, pp. 1022–1025.

Zrenner, E., Besch, D., Bartz-Schmidt, K. U., Gabel, F. G. V.-P., Kuttenkeuler, C., Sachs, H., Sailer, H. and Wilke, B. W. R. (2006). The active subretinal implant: 10 years of development to clinical application, in *104th DOG Annual Meeting*, p. online.

Zrenner, E., Miliczek, K., Gabel, V., Graf, H., E.Guenther, Haemmerle, H., Hoefflinger, B., Kohler, K., Nisch, W., Schubert, M., Stett, A. and Weiss, S. (1998). The development of subretinal microphotodiodes for replacement of degenerated photoreceptors, *Ophthalmic Research* **30**, 3, pp. 269–280.

Index

absolute refractory period, 114
accommodation, 30
action potentials, 37
Address Event Representation
  (AER), 218, 220
after-potential, 131–133, 135
age-related macular degeneration
  (AMD), 2
amacrine cells, 27, 34, 37, 43
  A17, 43
  AII, 43
  sustained, 44
  transient, 44
ambiguity, 194
Amplitude Shift Keying (ASK), 221
anterior chamber, 30
aqueous humour, 30
Artificial Silicon Retina (ASR), 202
autocorrelation function, 87, 136
  system input/output, 136
  white noise, 136
  z-transform, 136
autocovariance, 88
average neural response, 74
axon, 22
axonal terminals, 22

basal junctions, 36
Bayes' law, 98
Bayes' rule, 91, 141
Bernoulli distribution, 94
bilinear transform, 127, 128, 135, 150

Binary Phase Shift Keying (BPSK),
  221
binomial distribution, 94
Bio-inspired Visual Encoder, 212
BioElectronic Vision, 2
bioelectronic vision system, 5, 7, 199,
  207, 210, 225
bipolar cells, 27, 34, 37, 38, 40, 41, 43
  midget, 41
  OFF-type, 40
  ON-type, 39, 43
blindness, 1
  blindness distribution, 4
  profound blindness, 2, 11
  profoundly blind, 5
boxcar filter, 78
bursts of spikes, 178

cataract, 2
center surround organization, 40
central retina, 35
charged couple device camera, 201
chemical synapses, 23, 26
chloride ion (Cl^-), 24
ciliary body, 30
coefficient of variation (C_V), 99
conditional probability distribution,
  91
cone bipolar cells, 35
cone pedicle, 36
cones, 31, 33, 35–40, 43
  blue cones, 42

foveal cones, 32
S-cones, 37
connexin, 27
connexon, 27
continuous convolution, 78
contrast gain control (CGC), 50, 125, 128
convolution, 78, 123, 152
　continuous, 78
　discrete, 78, 152
cornea, 30
corneal opacity, 3
cortical neuroprostheses, 16, 199, 208, 210
cortical neuroprosthesis, 9–11, 15, 16, 211, 228, 229
cortical visual neuroprostheses, 14
Cortical Visual Neuroprosthesis for the Blind (CORTIVIS), 199, 210, 228
coupling transformer, 222, 228
crystalline, 30
cumulative density function (CDF), 143, 145

demodulator, 222
dendrite, 22
depolarization, 27
depolarizing, 39
deterministic model, 114, 124, 189
　assessment, 189
　computational implementation, 126
diabetic retinopathy, 4
difference of Gaussians, 125
difference of Gaussians (Dog), 125–127, 130
digital-to-analog converter (DAC), 222
Dirac delta function, 71, 72, 80
　properties, 71
discrete convolution, 78
discrete Heaviside unit step, 80
distance function, 155
distorted sinusoidal functions, 123, 130, 133
Dobelle, 208, 210, 228

dura mater, 30

Early Layers, 212, 214, 216, 217, 225
edit distance, 166
electrical synapses, 23, 27
electrode impedance, 222
electrodes, 8, 9, 14, 16, 61, 188, 201,
　see microelectrodes, 202, 203, 207, 208, 210
　penetrating, 14, 16
　surface, 14
epiretinal implant, 12
epiretinal neuroprosthesis, 10
ergodic process, 157
Euclidean distance, 171
expectation operator (E), 157
expected value, 96
experimental neuroscience, 57
extracellular recordings, 60, 188
eye, 5, 7, 14, 21, 28–30
　optical system, 1

Fano factor (F), 97
Field Programable Gate Array (FPGA), 224, 225
firing events, 178
firing rate, 51, 75–77, 79, 81, 86, 89, 93, 102, 104, 105
　average firing rate, 60, 76, 82, 86
　constant firing rate, 103, 104
　discrete firing rate, 82
　spike-count firing rate, 76, 77
　time-varying firing rate, 104
　variable firing rate, 104
firing rate metrics, 156
First In First Out (FIFO), 219
Forward Transmitter, 222, 229
Fourier transform, 151
fovea, 30–33, 35, 41
　perifovea, 36
fovea:parofovea, 35
frequency domain, see Laplace transform
frequency response, 151
　amplitude in dB, 151
　phase, 151

polar form, 151
Frequency Shift Keying (FSK), 221
full-field stimuli, 188
functional model, 53
functional models, 7, 113, 114, 123

Gabor function, 65
ganglion cells, see retinal ganglion
   cells, 33, 35, 37, 40, 58
   midget, 41
   OFF-type, 40
   ON-type, 39–41
gap junctions, 23, 27
Gaussian function, 66, 79
glaucoma, 2
glial cells, 23
glutamate, 27, 38

half-wave rectifier, 207, 222
Heaviside unit step function, 79, 121
homogeneous Poisson processes, 93
homunculus, 59
horizontal cells, 34, 36, 40
hyperpolarization, 27, 39

impulse response, 136, 137, 151, 152, 213
inhomogeneous Poisson process, 93
inner nuclear layer, 34
integral
   backward approximation, 151
   Euler approximation, 122, 149
   trapezoidal approximation, 150
integrate-and-fire model, 115
interspike interval (ISI) metric, 160
interspike interval distribution, 104
   homogeneous Poisson model, 98, 100
   inhomogeneous Poisson model, 103
interspike interval metric, 166
interspike interval probability
   inhomogeneous Poisson model, 103
interspike time interval, 98, 99, 120, 166, 168
intracellular recordings, 60

intracortical microstimulation, 210, 228
Intraocular Retinal Prosthesis (IRP), 200, 205
ion channels, 24, 26, 27, 53
ions
   chloride (Cl^-), 24
   organic ions (A^-), 24
   potassium (K^+), 24
   sodium (Na^+), 24
iris, 29, 30

joint probability distribution, 91

Laplace transform, 127, 128, 135, 146, 147, 151
   s-domain, 127
   region of convergence (ROC), 146
   s-domain, 150
Laplacian of Gaussian (LoG), 52, 127
lateral geniculate nucleus (LGN), 47
lateral geniculate nucleus (LGN), 46–48, 130
leaky integrate-and-fire model, 118
   computational implementation, 121
lens, 30
Levenshtein distance, 166
Load-Shift Keying (LSK), 226
look-up-table, 216
low-pass filter, 121, 128

macula lutea, 36
marginal probability distribution, 91
mean (statistics), 96
mean squared error (MSE), 156
mesopic vision, 31
metric, 155
microelectrode array, 14, 16, 69, 201, 210, 211, 228
Microelectrode Stimulator, 213, 221, 222
microelectrodes, see electrodes, 200, 210, 217, 218, 223
MicroPhotoDiode Array (MPDA), 202
minimum-phase system, 137

models
  deterministic, 114, 124
  functional, 113, 114, 123
  integrate-and-fire, 115
  leaky integrate-and-fire, 118
  rate-code, 113
  stochastic, 114, 129
  structural, 113
  time-code, 113
  white noise, 138
modulator, 222
moment generating function, 96
Monte-Carlo method, 159

neural code, 52
neural response function, 71, 73, 85
neuromorphic encoder, 211, 224, 228
Neuromorphic Pulse Coding, 217, 218
neuron, 22
neuron soma, 22
neuronal membrane, 23, 24, 38, 53, 115
  capacity, 115
  dynamics, 119
  potential, 115, 116, 119, 121
  resistence, 115
neuroprosthesis
  cortical, 5
  retinal, 5
neurotransmitter, 23, 26, 27
normalized mean squared error (NMSE), 157

On-Off Keying (OOK), 226
onchocerciasis, 4
optic nerve, 5, 11, 14, 33, 35, 37, 45, 46, 49, 50, 202
  neuroprostheses, 11
optic nerve neuroprostheses, 11
optic tract, 45
ora serrata, 32
organic ions (A^-), 24
orthonormal functions, 130
orthonormal vectors, 134
orthonormalization, 134
  Gram-Schmidt procedure, 133

outer nuclear layer, 34

parofovea, 35
partial refractory period, 114
pathways (M and P), 45, 48
percent-Variance-Accounted-For (%VAF), 158
peri-stimulus time histogram (PSTH), 82, 159, 180
perifovea, 36
phosphenes, 8, 9, 202, 208, 209, 227, 228
photodiodes, 13, 201, 202, 204, 207, 226
photopic vision, 31
photoreceptors, 8, 27, 31, 33, 37, 38, 43, 50, 202, 227
  cones, 31
  rods, 31
pia mater, 48
pial surface, 208
point process, 92
Poisson distribution, 96
  Fano factor, 97
Poisson process, 93, 97, 100, 104, 174
  Fano factor, 97
  homogeneous, 93, 98, 102, 173
  inhomogeneous, 93, 102
  spike train as a, 104
Poisson-like process, 140
post-synaptic neuron, 22
posterior chamber, 30
potassium ion (K^+), 24
power amplifier, 222
  Class-E, 222
pre-synaptic neuron, 22
pre-synaptic terminals, 22
pretectum, 46
prewhitening, 136, 137
probability density, 92, 95
  homogeneous Poisson spike train, 98
  interspike interval, 98
  moment generating function, 96
probability density function
  moment generating function, 97

probability mass function, 94
pupil, 29

radio frequency (RF) link, 221, 224
random access memory, 214, 217, 220
rate code, 51, 156
rate coding, 60
rate-code models, 113
read only memory, 214
receptive field, 37, 45, 86, 87
rectangular window, 78
refractory period, 26, 105, 114, 117, 118, 120, 131, 132
   absolute, 105, 114
   relative, 105, 114
region of convergence (ROC), 146, 148
retina, 1, 5, 10, 11, 21, 30, 31, 50
   central retina, 35
   deterministic model, 124
   functional models, 7
   retinal diseases, 2
   stochastic model, 129
   structural models, 7
   white noise model, 138
Retina Implant (RI), 200
retina model, 8, 14, 21, 53, 59, 113, 124, 138, 193, 212
retina models
   assessment, 188
   tuning, 188
retina nuclear layers, 31
retina plexiform layers, 31
retinal ganglion cell
   OFF-type, 84, 143, 145
   ON-OFF-type, 124
   ON-type, 84, 124, 143, 145
retinal ganglion cells, 12, see ganglion cells, 46, 47, 52, 87
retinal neuroprostheses, 11, 12, 15, 204
   epiretinal, 199, 200, 204
   subretinal, 199, 202, 204
retinal prostheses, 199
retinitis pigmentosa, 4, 13

retinotopic organization, 10, 49, 54, 200, 208
reverse correlation function, 83, 86
RF transformer, 222
rod spherule, 36
rods, 31, 36–38, 43

s-domain, see Laplace transform, 150
sclera, 30
scotopic vision, 31
Second Sight Medical Products, 200
serial communication link, 221
sigmoidal function, 143
sigmoidal nonlinearity, 145
signal flow graph, 214
sodium ion (Na^+), 24
spectral density, 136
spike
   bursts of, 178
   deletion, 162
   insertion, 162
   shift, 162
spike count distribution
   homogeneous Poisson model, 94
spike distance metric, 160
spike events metric, 177, 187
spike events metrics, 156
spike generation, 103
   integrate-and-fire method, 105
   Poisson process, 104
spike generator, 129, 138, 140
spike time metric, 160, 161, 164, 172
spike train, 58–60, 71
spike train autocorrelation function, 87, 88
spike train cross-correlation function, 88
spike train distance metric, 170, 171, 176
spike train metrics, 156
spike triggered average (STA), 83–87, 131
spike triggered covariance, 89, 90
spike-count distribution
   inhomogeneous Poisson model, 102
spike-count rate, 73

spike-triggered covariance, 89
spikes, 37, 58, 223
stimulus current, 115, 118, 120, 121, 123
stochastic model, 114, 129, 191
   assessment, 191
   computational implementation, 133
structural model, 115
structural models, 7, 113
subretinal implant, 12
superior colliculus, 46
synapse, 22, 26
   chemical, 23, 26
   electrical, 23, 27
synapse cleft, 27
synaptic cleft, 22
synaptic plasticity, 23
synaptic potential, 27
system function, 152

telodendria, 37
test data set, 188
time code, 51
time coding, 60
time-code models, 113
time-dependent firing rate, 75
trachoma, 4
training data set, 188
transmitter substance, 23
Tustin's method, see bilinear transform

unit impulse, 72
unit step function, 79
unit step sequence, 80

variability, 194
variance, 96
variance operator, 159
visual acuity, 204
   normal, 204
   sufficent, 204
visual channels, 39
   OFF-type, 39
   ON-type, 39

visual cortex, 5, 7, 8, 10–12, 14, 15, 21, 45, 48, 54, 203, 208, 210, 213, 223
   receptive field, 48
   visuotopic organization, 48
visual neuroprostheses
   cortical, 227
   cortical , 208
visual neuroprosthesis, 5, 17
   classes, 5, 7
   cortical, 9, 14, 15, 211, 212
   interfacing, 10
   research, 5
   retinal, 9, 12
visual stimulation, 69
visual stimuli, 61, 62, 65–67, 115
   non-uniform, 62
   ON-OFF, 62
   spatially uniform, 62
   stochastic, 66
visuotopic, 45
   mapping, 8, 47, 48
   organization, 8, 45, 46, 48
vitreous chamber, 30
vitreous humour, 30

white matter, 48
white noise, 87, 136, 137
white noise analysis, 138
white noise model, 138, 191
   assessment, 191
   computational implementation, 143
white noise stimulus, 66

z-domain, see z-transform, 150
z-transform, 127, 135, 148, 152
   inverse, 137
   region of convergence (ROC), 148
   z-domain, 127, 150
zonule fibers, 30